Technology and transformation in the American electric utility industry

Technology and transformation in the American electric utility industry

RICHARD F. HIRSH
Associate Professor, Department of History
Virginia Polytechnic Institute and State University
Blacksburg, Virginia

The right of the
University of Cambridge
to print and sell
all manner of books
was granted by
Henry VIII in 1534.
The University has printed
and published continuously
since 1584.

CAMBRIDGE UNIVERSITY PRESS
Cambridge
London New York Port Chester
Melbourne Sydney

CAMBRIDGE UNIVERSITY PRESS
Cambridge, New York, Melbourne, Madrid, Cape Town, Singapore, São Paulo, Delhi

Cambridge University Press
The Edinburgh Building, Cambridge CB2 8RU, UK

Published in the United States of America by Cambridge University Press, New York

www.cambridge.org
Information on this title: www.cambridge.org/9780521364782

First published 1989
First paperback edition 2002

A catalogue record for this publication is available from the British Library

Library of Congress Cataloguing in Publication data
Hirsh, Richard F.
Technology and transformation in the American electric utility
industry / Richard F. Hirsh.
p. cm.
Bibliography: p.
Includes index.
ISBN 0 521 36478 7
1. Electric utilities – United States – Technological innovations.
I. Title.
HD9685.U5H57 1989
333.79′32′0973–dc20 89-32978 CIP

ISBN 978-0-521-36478-2 hardback
ISBN 978-0-521-52471-1 paperback

Transferred to digital printing 2009

Contents

Preface

This book deals with technological stagnation and how it contributes to industrial decline. Focusing on the electric utility industry, the treatise offers a novel interpretation for the industry's woes: it argues that a long and successful history of managing a conventional technology set the stage for the industry's deterioration in the late 1960s and 1970s. After improving steadily for decades, the technology that brought unequalled productivity growth to the industry appeared to stall, making it impossible to mitigate the difficult economic and regulatory assaults of the 1970s. Unfortunately, most managers did not recognize (or did not want to believe) the severity of technological problems, and they dealt instead with financial and public relations issues that appeared more controllable. Partly as a result, the industry found itself in the 1980s challenged by the prospects of deregulation and restructuring.

In offering this view of the industry's problems, this book differs markedly from other works in history or business policy. This difference should also make the book of interest to several audiences. To historians, for example, this study provides a contrasting perspective to those that examine undaunted technological progress as a central theme in American history. Americans appear to have taken as a fundamental belief – even a faith – that technology always improves and makes their standard of living among the world's highest. Though impossible to quantify its contribution precisely, technological innovation is credited for much of America's tremendous increase in productivity during the last century. Advances in power machinery, transportation, and communications – to name a few – have increased the amount and quality of goods and services enjoyed by people.[1] The natural assumption made after examining this orthodox history is that technological progress will continue unabated and that it will enable further material advances in the future. It was a view accepted by engineers, managers, the general public, and even historians.[2]

Of course, progress does not continue forever. But rarely is this aspect of technological development discussed in the scholarly literature. In the few instances in which limits to technical advance are examined, they serve as the backdrop against which technological revolutions and other improvements are viewed. For example, in his rightly acclaimed book, *The Origins of the Turbojet Revolution*,[3] Edward Constant considers how scientific knowledge can offer a means by which stagnation is avoided and progress continued. His concept of "presumptive anomaly" helps explain how engineers employed scientific analyses to "predict" the most rapid speed that could be attained by propeller-driven aircraft. The studies spurred attempts to invent a different propulsion system – the jet engine – for overcoming this limit.

But technologies do not always undergo conveniently timed revolutions. Besides the electric utility industry, the automotive, steel, petrochemical refining, and rubber industries have watched productivity growth rates decay partly because they lost their ability to produce incremental improvements in existing process technologies or to develop revolutionary technologies as substitutes.[4] Even the "high tech" telecommunications industry has reached a point where new transmission technologies offer few cost advantages for the entire network.[5] In other words, the nature of technological or business enterprises does not necessarily guarantee that innovations will occur when they are needed most. Put still another way, not all technological problems are soluble. While some academics who deal with economic history believe that a free and efficient market exists for technological "progress" and that technological choices can be explained simply in neoclassical market terms, this may not always be the case. Some industries may have to wait years for needed technical advances. Others may be radically transformed before they arrive.

For readers interested in the history of business strategy, this book has policy implications simply because it focuses on a socially and economically significant industry. The electric utility industry provides the most versatile and desirable form of energy available. Its inability to produce abundant electricity at decreasing unit costs (as it had before the late 1960s) affects millions of people who developed a lifestyle dependent on this energy source. Cheap and available electricity, in other words, contributed to the material way of life that helped make Americans the "people of plenty."[6] Electricity's loss of these features has therefore had a major impact on the way people live. And because electricity propels hundreds of industries, trends in the utility industry are reflected in others. When the industry's productivity declines, the entire economy feels the repercussions.

Perhaps more importantly, the book examines one of the central problems in managing technology, namely, the difficulty of projecting accurately the direction and magnitude of future technological change. Several studies have focused on technological discontinuities and on the revolutionary innovations that created them.[7] This book, however, probes the unfamiliar but equally critical phenomenon of technological stability. It should therefore have practical and theoretical significance, providing a basis for business policy implementation by extending discussions on the creation of technological knowledge. While suggesting that technological stagnation need not be a permanent condition and that "rejuvenation" may be possible, the book's greatest use may be to sensitize business managers to technological standstill and to ways it can be avoided.

Students of business management may discover other interesting discussions as well. For example, the book offers a case example of how professional traditions, culture, and history affected the way technically trained managers directed a large technology-based industry. When running electric utilities, engineers often made decisions based on considerations of technological sophistication rather than on "simple" financial concerns such as profitability. Economic considerations, in other words, did not always rank prominently in the minds of engineer-managers, whereas technological prowess (as evidenced by deployment of the biggest and most technically efficient equipment) sometimes did. In short, utility managers often pursued goals that differed from those sought by accountants and other people who had more traditional backgrounds in business and finance.

Finally, the book suggests some business "lessons" for restructuring a technology-based industry. In early stages of a technological enterprise, a novel set of problems commonly gives rise to a "manager-entrepreneur" who can exploit features of a new technology and "rationalize" or make order of an industry. The solution to one problem, however, often leads to others in the form of financial constraints, later to be solved by "financier-entrepreneurs."[8] But the recent period in which technological stagnation has occurred in the utility industry calls for new people – certainly not the same type of people who organized a growing and incrementally improving technological system. Rather, they must be managers who can rerationalize the industry by comprehending technological barriers, social concerns, political realities, economic forces, and – perhaps most importantly – their own history.

* * *

My ideas about the technology employed in the electric utility industry began developing in 1979, when I served as chair of a citizens' committee overseeing rate "reform" for the Gainesville, Florida, municipal power system. The utility had just completed construction of a new power plant, and now the customers had to start paying for it. The elected politicians recognized an unpopular issue when they saw one, but they also had the wisdom to seek public input into the decision-making process. In my position as committee chair, I needed to learn quickly about power technology and utility economics. Then I realized the need to forge a consensus among committee members and public activists, some of whom desired low rates for commercial and industrial customers (to encourage business activity) and others who wanted high rates for all customers (to discourage waste). We eventually succeeded in developing an "inverted" block rate structure and an experimental time-of-day rate that all felt was reasonable and equitable. In gaining this education, I need to thank the patient staff of the Gainesville Regional Utilities system and Barney L. Capehart, professor of industrial and systems engineering at the University of Florida. Besides helping me understand the business, Barney provided a wonderful example of how engineers can solve social problems without reverting to technological "fixes."

Virginia Polytechnic Institute and State University (Virginia Tech) served as the next breeding ground for ideas. Through grants offered by the university's Center for the Study of Science in Society and the Virginia Center for Coal and Energy Research, I had the opportunity to design novel rate structures with physicist Samuel P. Bowen and learn more about the utility industry. Other Virginia Tech colleagues who helped me in formulating ideas and who read portions of my manuscript included Arthur L. Donovan, Rachel Laudan, Harold C. Livesay, Gary L. Downey, Saifur Rahman, Marjorie J. Norton, and Albert E. Moyer, the last of whom often coached me on style and content as we jogged along Blacksburg's wooded running trails.

A happy and unexpected course of events next brought me to the Harvard Business School, where I served for two years as a research fellow with Richard S. Rosenbloom. Through Dick's special form of tutelage, I learned much about the business management of technology. I know this book would have taken a very different form had I not become sensitized to business issues, and I am grateful to Dick and several of his colleagues at the "B-School" who generously helped me. These people include Thomas K. McCraw (who first brought my name to the attention of others at the school), Alfred D. Chandler, Jr., Richard K. Vietor, David A. Garvin, Robert H. Hayes, Roger E. Bohn, George C. Lodge, E. Raymond Corey, Robert D. Cuff, Kenneth A. Merchant, and Leslie R. Porter (another running partner who offered an abundance of good ideas).

While in Boston, I received an invitation from Professor Earl R. Mac-

Cormac of Davidson College in North Carolina to participate in a conference concerning electric power technology and values. The meeting with Earl began a solid professional and personal relationship that has benefited both of us. We bounced around ideas and tested new concepts, often while bouncing around a rubber ball in the squash court. (Earl usually beat me!) I gratefully acknowledge his enthusiasm for my work and his careful reading of all too many versions of this manuscript.

Thanks are due to many other people at various institutions. They include Warren D. Devine (Oak Ridge Associated Universities); David R. Nevius (North American Electric Reliability Council); Thomas J. Grahame (Department of Energy); Ronald R. Kline, Anne C. Benson, and Joyce Bedi (Institute of Electrical and Electronic Engineers); Spencer W. Weart (American Institute of Physics); Sam H. Schurr (Electric Power Research Institute); David K. Smith (Middlebury College); Herman Koenig (Michigan State University); George Wise (General Electric Company); and the staff of the Edison Electric Institute Library in Washington, DC. Extremely useful comments and suggestions also came from some of the leading members of the history of technology community, such as Thomas P. Hughes, Eugene S. Ferguson, Edward W. Constant, II, and Alex Roland. At the same time, I owe much to Ronald J. Overmann and Margaret W. Rossiter of the National Science Foundation and David E. Wright of the National Endowment for the Humanities, who encouraged my efforts and helped pay for an academic year of pure research. I also gratefully appreciate the time and effort expended by the managers, engineers, and regulatory officials who talked with me and who offered valuable insights into the history and management of the utility industry. (See the Bibliographic Note.) These people provided the source material from which I wove together many of my arguments and conclusions.

Finally, I owe much to my parents, sister, and grandmother for their intellectual and emotional support throughout the years. As the first academic in the family, I probably posed an enigma to them, since I worked in a strange field that crossed disciplinary boundaries (history and management of technology) and that was undoubtedly difficult for them to explain to friends and relatives. (If I wanted to be a professor, why couldn't I work in economics or physics – fields that people can relate to?) And to my beautiful new bride, Margene (who turned out to be my greatest "discovery" when I was on research leave at Harvard), I offer my thanks for abundant patience and understanding as I pursued this project. With good, common sense, she ordered me to stop building a room in our basement and complete my manuscript. She also gave me the "space" and comfort I needed, especially when working on one of those especially troublesome paragraphs or chapters. To her and the rest of my family – my original and newly extended family – I dedicate this book.

Acknowledgment of financial support

This material is based upon work supported by the National Science Foundation under grant number SES-8308407. Any opinions, findings, and conclusions or recommendations expressed in this publication are those of the author and do not necessarily reflect the views of the National Science Foundation. In addition, this publication has been supported by the National Endowment for the Humanities, a federal agency that supports the study of such fields as history, philosophy, literature, and the languages, under grant number RH-20539–84. I am extremely grateful for this assistance.

Introduction

In 1965, electric utility managers celebrated the eighty-third year of their industry's existence. No one held any "jubilee" festivities for this uneven anniversary, but signs of pride, confidence, and vitality could be seen everywhere: managers justifiably rejoiced as their power-generating technology recorded new heights in technical performance, contributing to the industry's unequalled productivity growth rate since the beginning of the century. They also congratulated themselves for managing a technology that supplied increasing amounts of electricity at declining unit prices, providing for higher standards of living during a period of general price inflation. Meanwhile, utility executives watched happily as investors bid up the share prices of their companies to new post-Depression highs, reflecting the view that previous trends in technology and business management would continue unabated.

By 1975, however, many of the same utility managers lamented their industry's condition. Instead of continued improvement, electric power technology appeared to have reached barriers that could not be breached. As a result, productivity gains disappeared, and the industry became susceptible to the same economic forces that disabled the overall economy. As the industry turned away from a pattern of declining unit costs, regulators abandoned their permissive role and became more activist, trying to represent cost-conscious consumers who ceased to view power technology as safe and benign. At the same time, utility managers encountered culture shock as they discovered that trends in electricity consumption had reversed themselves, and that "growth" no longer meant improved economic well-being for their companies or customers. Finally, investors forsook the electric utility industry as some firms approached uncomfortably close to bankruptcy. In short, the electric utility industry had been radically transformed in just ten years.

This book details the transformation of the electric utility industry. It focuses on the importance of technological progress in the industry's history and the business management principles that evolved to take advantage of improved hardware. But this book does not tell a success story. While providing a background glimpse of early accomplishments, it argues that the electric *utility* industry (which must be distinguished from the electrical equipment *manufacturing* industry) underwent a fundamental reorientation when the basic generating technology reached a pair of performance plateaus. Crippling the industry's productivity growth pattern, these consisted of barriers to thermal-efficiency improvements and to increases in scale economies. Experienced chiefly *before* the 1973 oil embargo, these limits contributed to the end of electricity's traditional features of cheapness and consistent availability. By concentrating on fundamental technological problems, this book therefore challenges the commonly held assertion that the industry's predicament stemmed *exclusively* from disruptions in energy supplies, financial market difficulties, environmentalism, inflation, and overzealous regulation.[1] Though not discounting these serious problems, this book simply argues that "traditional" studies do not paint a complete portrait. Inflation, for example, dogged the utility industry for decades, but it only became a heightened concern when manufacturers could no longer deliver new productivity-enhancing technology to mitigate it. In short, improving technology had always been a primary contributor to the industry's success and high productivity growth rate. When the technology reached apparent limits to improvement, it exacerbated an already decaying financial, economic, and regulatory situation.

To explain the causes of technological stagnation and decline in the electric utility industry, this book introduces the concept of "technological stasis." Stasis is the cessation of technical advances in an industrial process technology. Incremental improvements no longer are made, and the technology appears to have reached its limits. Stasis is not the same as what some people call technological "maturity," though it is related. A mature technology, according to some definitions, is one in which the basic design components of a process technology (or the products it creates) are well-defined. In the utility industry, for example, the design and successful use of steam turbines in the early 1900s set the agenda for further innovations.

Though "mature" as early as the 1920s or 1930s, power technology advanced in small, incremental steps for the next several decades. But during the 1960s and 1970s, barriers to improvement emerged in thermal efficiency and economies of scale. Now even the slow but steady progress ended, leading to industrial deterioration. Stasis therefore describes a condition that occurs in a technology that has already matured.

Stasis comprehends more than a hardware problem, however. It constitutes a technical condition that occurs within a social *system* of engineers, business managers, regulators, financiers, and the general public. Each set of participants (or "stakeholders" – people who have a direct interest in the operations of utilities) has different goals and agendas, and when they conflict, they can make a technology appear moribund. For the first half of the twentieth century, the engineer-managers of the equipment manufacturing firms and utilities developed technology that served all participants well. During the 1960s and 1970s, however, some players (utilities and manufacturers) tried unsuccessfully to speed up development of large-scale technology while others (consumers and regulators) began to distrust the actions of elitist technical managers. The resulting conflict exacerbated technical decay and seriously affected the industry's more obvious financial and regulatory woes. In short, this book uses the concept of technological stasis as a way to emphasize the social dimension of technical development. As such, the book offers a "sociotechnical" explanation for the recent decline of the electric utility industry. (See Appendix A for a more detailed discussion of stasis within the context of technology life-cycle models.)

As upsetting as it was, stasis did not occur throughout the world's electric utility systems. Rather, it remained an American phenomenon. Several factors account for this localization. For one, the United States constituted the world's largest market for power equipment – in 1969, it contained 43% of the noncommunist world's installed capacity – and it traditionally produced the greatest demand for new technology.[2] And because of an unusual form of competition between utilities (described in Chapter 5), American companies sought technology that continuously offered greater fuel efficiency and larger scale. If practical limits in technological advance were to be encountered, then they would show up first in the United States, where technically aggressive utility managers ordered large quantities of state-of-the-art equipment earlier than their counterparts in other countries.[3] In addition, the United States sported a decentralized and pluralistic utility industry consisting of hundreds of independent power companies, largely financed through free-market mechanisms and governed loosely by state and federal regulatory bodies. As a result, the American system could be affected by a variety of participants that contributed to the onset of stasis in a way that could not be easily duplicated in many other countries. For these reasons, the account that follows describes events occurring in the United States.

A few simple graphs will clarify the problem addressed in this book.

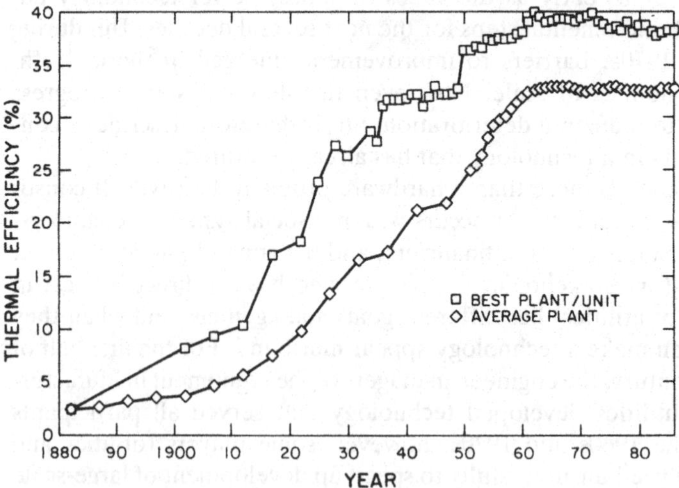

Figure 1. Thermal efficiency of generating units, 1880–1986. Thermal efficiency of power units increased gradually throughout the industry's history, plateauing in the 1960s. For the years after 1965, the data on the top curve relate to the most efficient *unit*. Before 1965, the data relate to the best plant – a combination of units. Data are from Federal Power Commission publications and annual reports of best thermal efficiency in *Power Engineering* magazine.

They also outline its basic themes about technological stasis. Consider, for instance, a graph of thermal efficiencies of power plants (Figure 1). Rising steadily throughout the first eighty years of the electric utility industry's history, the curve flattens out in the early 1960s and remains unimproved into the 1970s, showing that utilities could no longer economically coax more electricity out of a pound of coal or a barrel of oil.[4] Meanwhile, other graphs demonstrate that the capacity of new power units had also leveled off, this time in the 1970s (Figures 2, 3, and 4). Since the increasing output of units generally provided economies of scale that helped reduce unit costs, the flat curve in the 1970s meant bad news. Together, the end of thermal efficiency and scale improvements contributed to the reversal of a trend toward productivity improvements – improvements that previously made electric utilities the marvel of American industry.

These graphs do not necessarily prove a correlation between technological stagnation and industrial decay. But because they demonstrate that significant trends in the industry had begun to change well before 1973, they prompted an examination beyond the common interpretation of the electric utility industry's decline. That is, they encouraged a look beyond energy-supply distortions and economic, financial, and regulatory problems. This book is a result of that examination. Its first part, "Progress and Culture," provides the technical and social background for the utility indus-

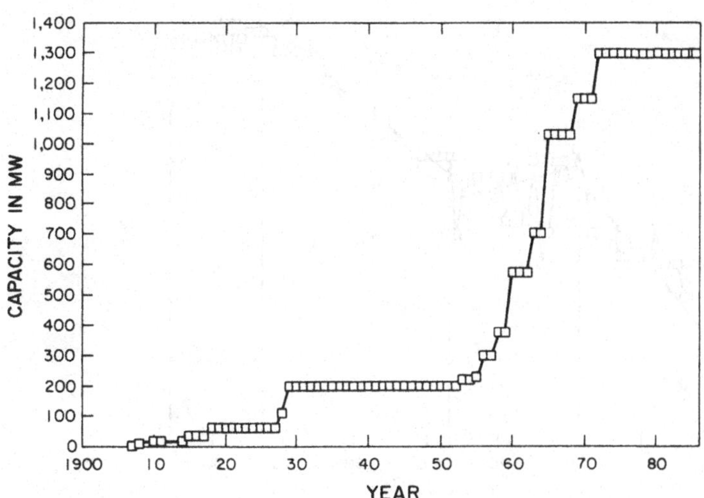

Figure 2. Maximum capacity of extant power units, 1900–86. The output of the largest steam-turbine generator grew dramatically in the period before the Great Depression and after World War II. A "unit" is defined as a "tandem-compound" or "cross-compound" turbine generator. Data are from U.S. Department of Energy, Generating Unit Reference File.

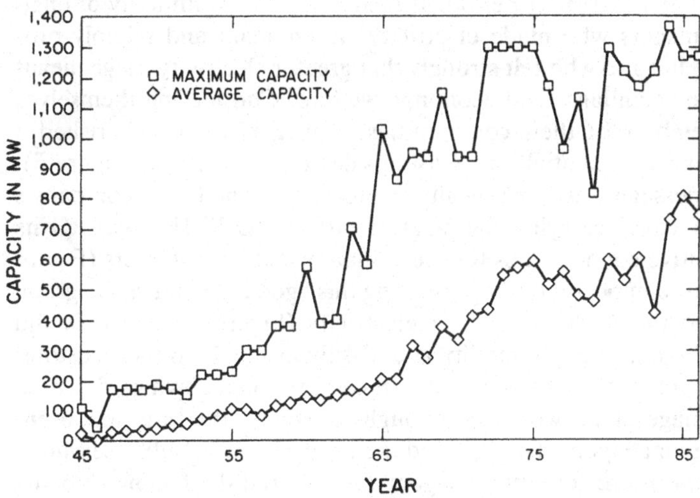

Figure 3. Maximum and average capacity of new units, 1945–86. The output of newly installed fossil-fueled and nuclear units increased until the early 1970s. Data are from U.S. Department of Energy, Generating Unit Reference File.

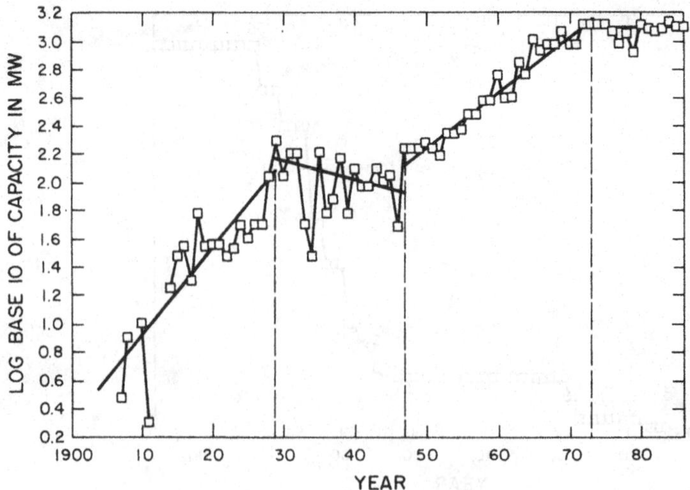

Figure 4. Largest unit installed by year. This logarithmic graph emphasizes how unit size increased exponentially before the Great Depression and after World War II. Two periods of stability existed from 1929 through the end of World War II and after 1973. Data are from U.S. Department of Energy, Generating Unit Reference File.

try in the post-World War II period. It focuses on the community of engineers and managers who made electricity an abundant and reliably produced commodity and who felt strongly that greater electricity usage meant increased living standards and economic welfare. Considering themselves important members of their communities, managers rightly cherished a graph recording exponentially increasing sales until the 1970s (Figure 5). The graph's message was symbolically reinforced by one utility company's annual report cover carrying the single word "GROWTH," with all its attendant positive connotations for the company and its customers (Figure 6). Perhaps more important in representing this "good" feeling is the graph that illustrated the declining cost of electricity (Figure 7). Not only did managers make a useful commodity abundantly available to their customers, they also did it in such a way that unit costs spiraled downward. These graphs and image alone would be enough to explain the satisfaction enjoyed by utility managers. They viewed themselves as true public servants.

But there was more. Utility managers also savored the feeling of being responsible stewards of technological progress. From the start of the industry, managers – who overwhelmingly had engineering backgrounds – used improving technology to raise thermal efficiency and to increase the scale of operations. To them, lower "heat rates" (a measure of greater efficiency) and the ever-increasing scale of turbine generators symbolized the techno-

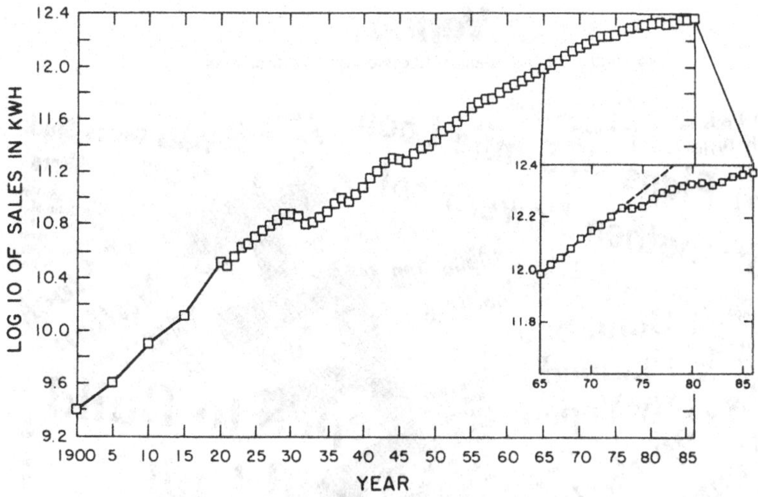

Figure 5. Sales of electricity, 1900–86. This logarithmic graph shows steady growth of electricity sales until 1973, when the Arab oil embargo and much higher energy costs spurred conservation efforts by users. Data are from Edison Electric Institute, *Edison Electric Institute Pocketbook of Electric Utility Industry Statistics*, 34th Ed. (New York: Edison Electric Institute, 1988.)

logical achievements that meant benefits to everyone (Figure 8). In short, these illustrations connote the exuberance of utility people who, for almost a century, understood and managed a complex technological system with little interference from "outsiders." They derived pleasure in knowing that their work contributed to the public good, and they developed a culture – a set of values, assumptions, and historical lessons – that became entrenched as they pursued further progress into the 1960s and beyond.

This book might be interesting in itself if it simply explained the development of a managerial culture that grew out of a specific set of historical experiences with technological systems. But such an exposition is merely prologue to a discussion about how this history affected management decisions concerning a stagnating technology. In the second part, "Stasis," the book argues that electric generating technology, which traditionally improved through small steps, no longer achieved many gains in the 1960s and 1970s. The text first explores the technical reasons for limits to progress in the industry. It continues with an explanation of how management practices in the utility and power-equipment *manufacturing* businesses contributed to the creation of barriers and how responses to stasis by *utility* managers did not help the situation. For example, the book argues that utility executives ignored some signals emerging from the graphs displaying stagnation because they continued to rely on successful, but

Vepco

1968 ANNUAL REPORT/VIRGINIA ELECTRIC AND POWER COMPANY

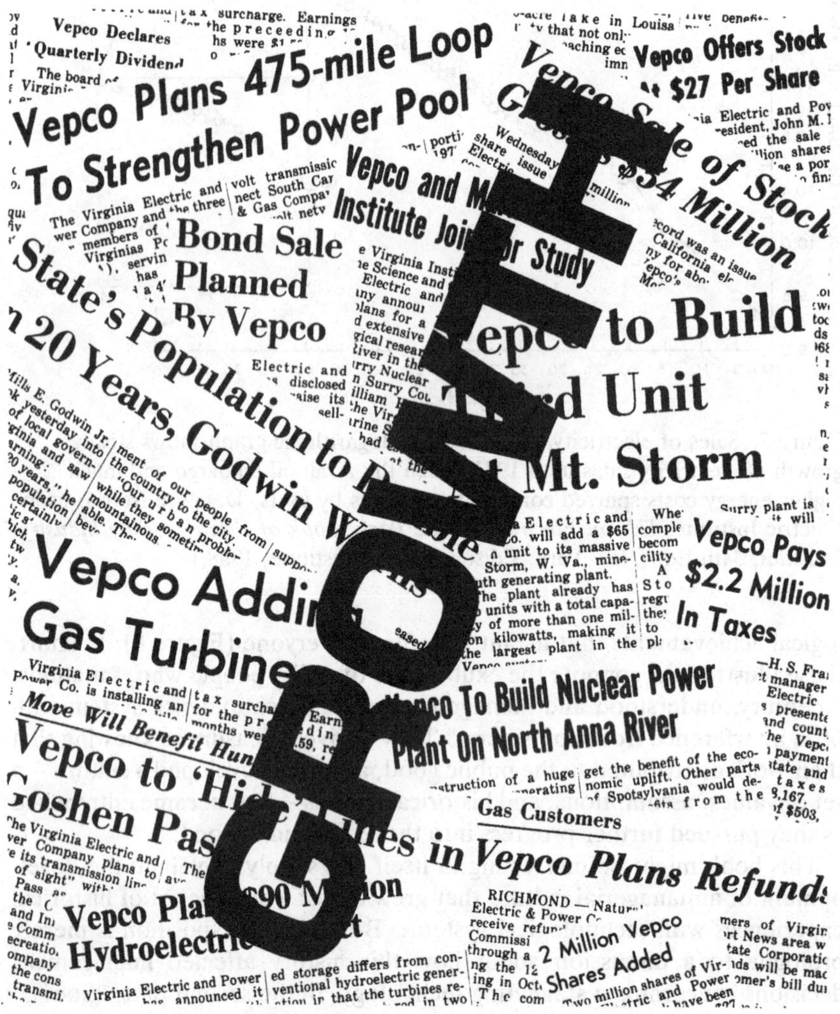

Figure 6. Cover of the 1968 *Annual Report* of the Virginia Electric and Power Company. The single word "Growth" implied good things to utility managers, stockholders, and customers in 1968. Reprinted with permission from Virginia Power Company.

now invalid, business assumptions and practices that no longer had sound bases in experience.

Exacerbating technical problems, the industry suffered after World War II from its inability to lure the most innovative engineering and manage-

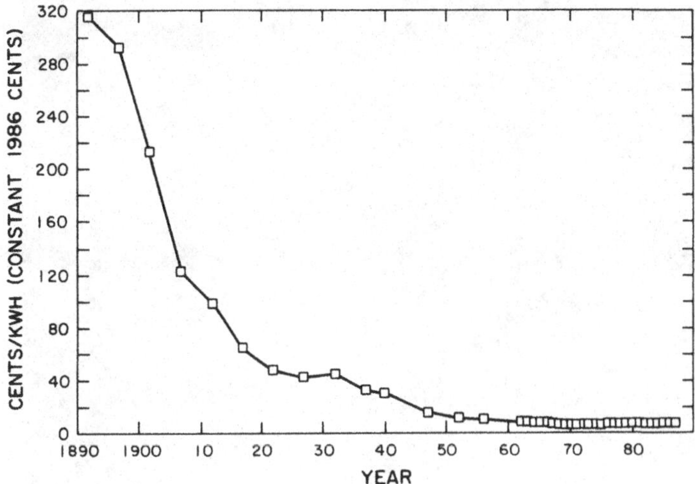

Figure 7. Average adjusted price of electricity. For residential service, the real (inflation-adjusted) price of electricity fell (in terms of 1986 cents) from over 300 cents per kilowatt-hour to about 7 cents per kilowatt-hour. Beginning in the late 1960s, the real cost of electricity ceased declining, so that the price in 1986 was about the same as it was in 1970. Data are from Edison Electric Institute, *Edison Electric Institute Pocketbook of Electric Utility Industry Statistics*, 30th Ed. (New York: Edison Electric Institute, 1984) and "1988 Statistical Report," *Electrical World* 202 (April 1988): 61.

ment talent available. Once viewed as exciting and "high tech," the electric utility industry lost recruiting battles to the electronics and aerospace industries, which carried with them the aura of exciting technological frontiers. By failing to attract the most competent individuals who may have anticipated stasis, the utility industry was left with people who would not challenge long-standing practices and assumptions that had become part of utility culture. They therefore allowed old assumptions to become even more entrenched, and they resisted change when a new business and technological environment emerged in the late 1960s. They also resisted efforts to cooperate in performing research and development on new forms of power technology, relying on their manufacturers to continue doing what appeared to be eminently successful in the past.

The third part, "Accommodating Stasis," describes basic strategies that some managers have inaugurated for dealing with an industry whose technology has reached stasis. As a first step in this direction, managers have been required to understand the values of each of the stakeholders in the power matrix and realize why conflicts arose. In the "good old days" before the 1960s, a consensus had been established among parties, and it accorded great power to utility executives to pursue technological developments as

Figure 8. The author standing next to a 1,145-MW-turbine-generator set. Photograph by Frank W. Bliss.

they saw fit. But when consumers, for example, began viewing electric power technology as expensive, polluting, and dangerous, they influenced regulation and hindered managers' efforts to pursue traditional goals. As the consensus unravelled, some managers in the 1980s began to recognize that their values and business strategies must change too. For the short term at least, they realized that a new consensus could be built around managing the existing technological system and discouraging the need to install large amounts of new equipment.

For the longer-term future, however, new technology will be required to meet the demands of a growing country. One approach to develop the equipment has consisted of overcoming stasis and effecting further technical advances in conventional technology. (Indeed, one of the attractive features of the notion of stasis is that it leaves open the possibility of further technical improvements.) The chapter entitled "The Search for New Technology" therefore discusses attempts made to overcome stasis. But the new technology may not resemble what predominated during the first seventy years of this century. Instead, to achieve a new consensus, some players are developing small-scale technology, to be operated by unregulated firms that compete or collaborate with established utility companies. Though highly controversial among utility managers, this new approach may have the effect of rerationalizing the entire industry. It may also be a way for the industry to provide enough electricity at reasonable prices while also balancing the needs and values of various participants.

Finally, the book reiterates the importance of appreciating historical experience as a factor in running a business or industry. The conclusion suggests that in order to manage technological enterprises successfully, individuals should realize that a history of accomplishment can adversely affect management behavior in the future. It therefore cautions against reliance on past practices and advises that awareness of an industry's technical and cultural roots must be cultivated by today's managers. In short, the book argues that studies of the past should play a significant role in managing events in the future. Since views of current technological and business situations are molded by representations of the past, managers must use historical understanding to distinguish between unchallenged assumptions and wise practice. This lesson holds true as much for other businesses as for electric utility companies.

Part I
Progress and culture

"Live Better Electrically"

– *Electric utility marketing slogan, late 1950s and 1960s.*

1 Managerial and technological foundations

Technological stasis in the electric utility industry occurred after decades of progress. "Progress," of course, was not a term to be debated in academic forums. To utility managers, it had a simple, straightforward, and deterministic meaning; namely, it referred to the good things they did that continually got better: as technology improved through incremental gains and as sales of electricity increased, their companies grew to be more financially secure and profitable in a way that pleased managers and investors alike. Meanwhile, utility customers benefited because they received reliable service and more kilowatt-hours (kWh) of energy at declining unit costs.

This sense of progress emerged early in the utility industry's history. It derived from the development of fundamental business management techniques and related improvements in power technologies. While perhaps commonplace and even common sense today, these principles were not immediately obvious, even to such pioneers as Thomas Edison. Nevertheless, by the 1930s, the utility industry had firmly established a "grow-and-build" strategy as the basic management approach and had developed associated principles such as promotional pricing and holding companies to make it work. The strategy also depended on the development of large-scale technology that offered decreasing unit costs – something the newly developed steam turbine generator was able to provide. Together, the specific strategy and technology could be considered the "dominant design" of the electric utility industry.[1] In other words, through a unique political and social process, the early entrepreneurs chose the fundamental parameters and core elements that defined the American utility industry and distinguished it from those in other countries as well as from other possible permutations in the United States. Once created, the dominant design helped managers set priorities and goals as well as conduct their day-to-day activities.

Variety and first management principles

Variety of options characterized the early years of the electric utility industry just as it attends the building phase of almost all industries. Pioneers needed to rationalize an inchoate set of technological elements and management principles in order to create a system that could generate growth and profitability. As the first pioneer in the industry, Thomas Edison established a power company in New York City in 1882 for selling electricity to owners of his new incandescent light bulbs. While others had invented similar forms of light bulbs – the British often call Thomas Swann the bulb's inventor – Edison became widely acknowledged as the creator of the first successful *system* for producing, distributing, and using electric power.[2] Employing the terminology of historian Thomas P. Hughes, Edison was an "inventor-entrepreneur." He focused on the invention of a new technology, and he also established a business organization that could profit from it.

Edison's power station consisted of coal-fired boilers that produced steam for reciprocating steam engines – bigger and more sophisticated versions of machines resembling those built by James Watt a century earlier. Mechanical energy from the engines turned generators (basically magnets housed within coils of wire) to produce direct-current electricity. Passing through underground wires, the electricity found its way to users in homes and offices within an area of about one square mile in the Wall Street district of New York City (a good place for publicizing his new system to potential clients and investors). The system appeared to work well, becoming the model for franchised Edison Electric Light firms in various cities that bought technology from Edison's manufacturing companies.[3] Like Edison's New York City plant, these companies sold electricity only to small areas near the central station. For isolated customers, such as hotels or factories, Edison marketed self-contained engines and generating systems.

As with many businesses that introduced popular products, Edison's company met competition from other entrepreneurs. George Westinghouse and Elihu Thomson constituted the most formidable rivals, who experimented with various networks of direct current (DC), alternating current (AC), and mixed currents in attempts to establish the optimum power system.[4] Edison fought to preserve and expand his direct-current systems, but the advantages of alternating current for long-distance transmission eventually made AC the winner of the "war of the currents."[5] To survive the challenge, Edison's company merged in 1892 with the Thomson-Houston firm (an alternating-current champion) to become the General Electric Company.[6]

One central management issue arose quickly in the electricity supply business, namely, how to price the product so that increased sales and profits would result. But here was a strange product – a shapeless, invisible, and intangible form of energy. When Edison contemplated how to charge for electricity, he looked to his competition in the gas supply business for guidance. Accordingly, he established a "fixture rate" by which customers paid a flat rate for each bulb used. The rate did not differentiate between large and small users of electricity because, at the time, Edison had not yet invented meters for measuring consumption.[7] Nevertheless, after two years of losses, Edison's company made money as it sold electricity to wealthy homeowners and business customers who used the novel lights in window displays.[8]

As the new Edison Illuminating companies and competitors proliferated in several cities, the pricing issue became a major management concern. The basic problem consisted of why profits did not increase along with growing sales. As one utility manager reported in 1896, his company witnessed a deterioration of earnings with increased business. "To find that too much business was undermining the stability of the company and jeopardizing its success was startling," he commented. "It became evident that something was wrong."[9] According to another manager, Arthur Wright of Brighton, England, the problem existed because people still did not understand how to determine the costs of producing electricity. Without such an understanding, they could not develop rate structures that guaranteed profits.[10]

This same problem vexed Samuel Insull. The British-born secretary of Thomas Edison, Insull left his boss's employ in 1892 to become president of the small Chicago Edison Company, one of twenty electric companies in the city. When Insull took over the business, Chicago boasted a population of more than one million, but only 5,000 people had electric lights. While optimists suggested that as many as 25,000 Chicagoans might *ultimately* use electricity, Insull vowed to serve the entire population. But like other utility managers, Insull puzzled over the enigma of decreased earnings with increased investment.[11]

While companies foundered, the theoretical basis for rate making became established. The major conceptual advance came from British engineer John Hopkinson, who differentiated the costs of producing electricity into two categories: standing (or fixed) costs and running (or operating) costs. Standing costs related to land, buildings, and equipment that were independent of its use, whereas operating costs referred to the fuel, labor, and maintenance that depended on the amount of electricity being produced. In discussing these costs, Hopkinson referred to the example of a metropolitan railroad line that brought passengers into and out of cities at the beginning and end of the day, but which remained largely unused during the day's balance. By necessity, the lines required a carrying capacity equal to their greatest demand. This meant that companies made large investments in rolling stock, even though they used it for short periods only. Because the companies incurred high costs to supply this demand, Hopkinson argued that they should charge more during these periods.[12] In the electricity business, a similar situation existed. Utility companies needed to be capable of supplying the highest demand, even though customers infrequently called upon maximum capacity.

In order to improve returns on investment, Hopkinson employed the concept of "load factor," the ratio of average daily (or yearly) use of electricity to the maximum load sustained during the same period. If usage could be distributed throughout the day in order to raise the load factor, the unit cost of operation would be minimized as standing costs become shared by many customers. Put another way, because the standing costs are similar for systems that have a high and low load factor, "the cost of supplying electricity for 1,000 lamps for ten hours is very much less than ten times the cost of supplying the same 1,000 lamps for one hour, particularly if it is incumbent on the undertaker to be ready with a supply at any moment that is required."[13] He concluded that if companies obtained high load factors, the cost per unit of electricity would decline. "It is obvious that those whose use is long will find the electric light economical to themselves and that it will be profitable to the undertaker."[14]

Arthur Wright seconded and extended Hopkinson's conclusions. Besides differentiating between standing and running costs, Wright found that a chief factor in determining costs consisted of the maximum annual load that a power plant could provide.[15] For all practical purposes, he maintained, a customer's standing charge should be proportional to his heaviest load, which could be measured with a special device known as a "Wright demand meter." For maximizing the return on investment, Wright suggested that companies find customers whose maximum use of electricity occur at different times during the day, thus spreading out the load evenly. This concern for diversity (and the so-called "diversity factor") would prevent all customers from requesting large amounts of electricity simultaneously, thus avoiding the need to make large investments in infrequently used generating equipment. By promoting diversity, then, the utility company could serve all its customers more economically with less generating equipment.[16]

Buoyed by this knowledge gained from meeting with Wright, Insull sought new customers who collectively would provide high loads and good diversity factors. To do this, he tried to entice large industrial users, such as electric railroad companies and factories, which would provide loads during the day to balance nighttime residential use. But industrial customers could often afford to install their own power equipment, leaving customers who had low load factors to Insull. Realizing the need to capture the big daytime users, Insull offered competitive rates to them, making it uneconomic for them to produce their own power. As he pointed out to other utility managers in 1910,

> If you will bring your price down to a point where you can compel a manufacturer to shut down his private plant because he will save money by doing so; if you can compel the street railway to shut down its generating plant; if you can compel the city waterworks, whether privately or publicly owned, to shut down its power plant because of the price you quote – then you will begin to realize the possibilities of this business, and these possibilities may exceed your wildest dreams.[17]

The key management development, then, consisted of an understanding of the utility cost structure and the importance of load factor and diversity. Since companies could improve their returns on investment by attracting customers who boosted the load factor, managers learned that they could lower their own costs, expand service to new users, and offer lower rates to everyone. As Insull discovered, promotion of electricity to valued customers helped him reach his objectives. By employing this strategy, Insull consolidated his hold on the electricity supply business in Chicago and provided an example to other electricity entrepreneurs.[18]

Technology and the "grow-and-build" strategy

While Insull's strategy initially proved successful, it relied on more than an understanding of load factor and diversity. It depended as much (if not more) on improvements in technology that overcame limits inherent in the reciprocating steam engine. This power source, acceptable for many manufacturing applications, simply had become too large and unwieldy for electricity production. Some of its components became so huge that transportation and installation had become almost impossible. And its noise, smoke, and throbbing motions often caused nearby residents to complain – an early case of environmental opposition to electric power plants.[19] If growth in the electric utility industry were to continue, Insull would need to install multiple units of steam engines of limited output – a maximum of about 5,000 kilowatts (kW) each – and would not be able to take advantage of scale economies that could conceivably be gained by using more powerful machines.[20] In short, extant technology limited the advance of the electric power business. Perhaps, even with Insull's new management concepts, electricity would remain a luxury item.

Fortunately for Insull and his industry, England's Charles A. Parsons recognized the limits of the reciprocating engine and developed an engine in 1884 that operated on radically different principles. Instead of cyclic motion, the steam turbine produced rotary motion directly as high-pressure steam pushed against blades attached to a shaft, which was connected to a generator. In the eyes of engineers, the turbine exhibited greater "elegance" because of its design simplicity. Meanwhile, the machine occupied about one-tenth the space, weighed one-eighth as much, and cost one-third as much as a reciprocating engine of equivalent power while humming softly and without vibration noticeable outside the plant.[21]

The widespread use of turbines did not occur automatically, however. European companies began to employ them in capacities of up to 4,000 horsepower (almost 3,000 kW) in the first years of the twentieth century. But Insull wanted still more powerful machines to achieve scale economies and to pursue his goal of consolidation. Unfortunately, American manufacturers would not produce a larger machine and guarantee its performance, arguing that they simply did not have the technical ability to do so. But Insull persisted, and he convinced the General Electric Company to assume the manufacturing risks of building a 5,000-kW unit while his company took all the operating risks.[22] Operated first in 1903, the unit was followed by three others in 1905 and then replaced *en masse* in 1911 by 12,000-kW machines. Having demonstrated the viability of the new technology by these daring moves, Insull's company started a bandwagon effect that helped create a thriving turbine manufacturing business in the United States and the acceptance of the turbine as the power industry's prime mover.[23]

Insull's use of turbines went hand in hand with his adoption of alternating current and high-voltage transmission systems. Though his Chicago company had originally been devoted to Edison's direct-current system, Insull recognized how AC could help him achieve his goal of serving a large region. Unlike DC electricity, AC can be transformed to high voltages and transmitted over great distance with little power loss. An impressive demonstration of long-distance transmission occurred in 1895 when Buffalo, New York, drew power from generators at Niagara Falls, a distance of twenty-two miles. Several western cities also depended on water power (or "white coal") transmitted from mountain sites at increasingly high voltages.[24] Insull's AC transmission system employed the same basic principles, but it served a different purpose. By reaching out beyond Chicago's city limits with transmission lines and interconnecting with towns that had isolated power stations (or none at all), he could employ his larger and more efficient turbines and generators to provide a more diverse base of customers. While transmission lines more than doubled his investment cost per kilowatt of capacity in his first rural experiment between 1910 and 1912, the benefit of higher load and diversity factors compensated to such an extent that the total cost per kilowatt-hour declined by 60%.[25] During World War I, the idea of interconnection between different firms that had complementary load patterns became widespread as a way to improve overall load factors and increase energy efficiency. After the war, utility companies in several parts of the country pursued interconnection. As was becoming more widely known, interconnection led to lower production costs because small, inefficient, and isolated plants could be eliminated by using economical large plants. Greater reliability of service also resulted from employing a few larger power stations rather than a host of small ones. And because power systems no longer depended on individual plants, companies could lower reserve margins (and the cost of building extra plants), depending on other stations in the system to provide electricity if one were forced out of service.[26]

Though not articulated as such at the time, Insull had developed a business plan that would become known as the grow-and-build strategy. Based on the presumption that new technology would become more efficient and less costly (per unit of power output), the strategy simply encouraged the growth in customer usage of electricity so that utility companies would need to install new power units. Because the new units generally had greater output ratings and were more thermally efficient, providing economies of scale and lower operating costs, and because the cost of the new equipment would be spread over a greater number of users (and kilowatt-hours), the expense of providing electricity declined as utilities produced more of it. When the per-unit cost of increased output – known as the marginal cost – declined, prices would usually follow, thereby benefiting customers who used it in greater quantities.[27] The increased demand for the product would create a need for still more power units that offered greater scale economies and thermal efficiency. In a self-reinforcing manner, growth in usage, new construction, and lower costs would continue indefinitely, or so many managers believed.

The beauty of the grow-and-build strategy – and the reason why Insull deserves being called a "genius" in this case – is that the approach meshed the interests of utility companies, its investors, and equipment manufacturers with those of consumers. He demonstrated that promotion and load growth should be viewed as good because all participants in the electric power matrix benefited with increased business and lower costs. Through a propitious mix of technology and management, Insull made his Commonwealth Edison Company known as one of the world's greatest electric utilities.[28] In the process, he also popularized the grow-and-build strategy, which, by the 1920s, appeared to be the only possible and logical approach for running a utility company.

Regulation and holding companies

Before the new strategy could succeed on a widespread basis, however, two more organizational principles needed to be established. One was legalized monopoly control. Aware of the potential cost savings that could accrue from using new and larger generating technology, Insull sought to unify several small companies in the Chicago area, thus justifying purchase of the equipment in the first place. He also understood that with the huge capital outlays for generating plants and distribution facilities, it made little economic sense for more than one utility company to serve each community. If he had exclusive rights to produce power in a region, however, his monopoly company could generate electricity at the lowest possible price, and everyone would profit.[29]

Not everyone shared Insull's view of the way the power industry should be structured, however. Since the beginning of the central power station business in the 1880s, several city governments established utility companies (or bought out existing privately owned firms) to provide electricity in the same way as they provided necessities such as water and sewage services. Because municipal governments paid no dividends to shareholders and could obtain low-cost loans by issuing tax-exempt bonds, they often could produce electricity at lower cost than private companies. Civic reformers, trust-busting politicians, and muckraking journalists supported the public power movement during an era when "big business," as typified by John D. Rockefeller's Standard Oil Company, acquired negative connotations. By 1902, 815 municipal systems had already been formed in the United States, and their combined capacity accounted for 9.3% of the nation's total.[30] Supposedly insulated from business corruption, these public power systems constituted an alternative model for the utility industry.[31]

Within this increasingly hostile environment, Insull knew that he would need to garner popular acceptance in order to win monopoly status for his utility. Joined by a few other like-minded utility entrepreneurs such as Detroit Edison's Alex Dow,[32] but severely opposed by others, Insull advocated as early as 1898 that private companies make a bargain with state governments: in return for exclusive production and distribution rights within a region, the private utility would accept regulatory oversight by a governmental body. Though the political situation in Illinois militated against early approval of this idea, it caught on elsewhere. In 1907, New York, Massachusetts, and Wisconsin took the lead by establishing public service commissions to oversee the practices of electric utilities. Not coincidentally, these moves occurred just as the public power movement stalled during a financial panic that destroyed investors' faith in municipal bonds – the primary means used for financing city-owned systems.[33] Illinois eventually followed by creating its own regulatory body in 1914.[34] Patterned after regulatory commissions that already existed to deal with railroads and other natural monopolies, these (and later) bodies had two objectives: to assure the public that it would receive a socially and economically significant commodity – electricity – at reasonable rates and with reliable service; and to enable utility companies to earn a "reasonable" return on investments so they could produce electricity, construct new facilities, and, in general, maintain their financial integrity.[35]

Despite the best intentions of state governments, regulatory bodies generally did not receive the resources or staffing to make them strong, effective counterforces to the utilities. The bodies therefore constituted a relatively benign participant in the electric power matrix, legalizing monopoly control of utilities and permitting managers great control over the industry.[36] During a period of general price decreases in electricity, few consumers or other interveners (such as the gas utility companies) had much impact on state regulation of the power industry.

A final organizational principle that several utilities exploited came from outside Insull's firm. "Invented" by S. Z. Mitchell, a Naval Academy-trained engineer who worked with Edison's New York company, it consisted of the holding company, a means by which utility systems could acquire financial resources for expansion. As chief agent for Edison products in the northwestern states, Mitchell learned of the problems many small companies encountered in financing the purchase of equipment from manufacturers. Basically, the firms needed an extraordinary amount of money to finance purchases – as much as $4 to $6 of capital to produce $1 in annual gross revenue.[37] In 1890, the Thomson-Houston company (and later the General Electric Company) developed a means to help small utilities by accepting the companies' relatively unattractive bond securities as collateral for bond issues of a larger company (known as the United Electric Securities). Because of the larger firm's size and because of its relationship to the manufacturing company, it enjoyed some success in selling the "bundled" securities of the small utilities. But the 1893 panic saw many utility companies fail, making investments in their bonds unprofitable – even through the United Electric Securities. To provide greater safety, bond buyers wanted these firms to issue more common stock and less debt. With fewer accounts payable, the companies would not be as vulnerable when revenues fell during business downturns.[38]

Mitchell's innovation of the holding company resolved fears and provided new opportunities. As a "financier-entrepreneur" (in Hughes' terminology), Mitchell presided over the creation of the Electric Bond and Share Company in 1905. After purchasing or exchanging the unmarketable stocks and bonds of individual utilities, the firm issued its own securities. Meanwhile, the operating companies still issued bonds for about 60% of the companies' capitalization, with preferred and common stock taking up the remainder of the financial needs. The arrangement offered investors greater security through diversification, but it also meant that control of the smaller companies through common stock remained relatively inexpensive and could lead to high investment returns for the holding company.

But the relationship between the holding company and utility served more than just financial needs – and here was perhaps the greatest innovation at the time. The holding company also provided management and engineering services, thus helping utilities return to financial health after the panic. Large companies such as Insull's Commonwealth Edison could afford to provide these services itself, but most of the newly formed smaller utilities could not. By pooling resources and providing for a centralized resource of management and technical expertise, the holding company could offer these specialized services to new companies and help keep them operating efficiently and profitably. Later, the holding company organized transmission interconnections with other companies and oversaw their expansion.[39]

Overall, the holding company innovation impressed many interests. Investors liked them because holding companies held securities in diversified operating firms. If one utility fell on hard times, others in the portfolio could compensate. Meanwhile, investors had assurances that the operating utilities would be well managed through services provided by the parent company.[40] Utility companies benefited too by being able to raise money needed for what quickly became the country's most capital-intensive industry. Signifying the success of the system, new holding companies proliferated, peaking in 1924, when only twenty holding companies controlled 61% of the total generating capacity of commercial electric power plants.[41]

In many cases, however, expansion did not follow technological imperatives, with some managers abusing the system and exploiting the speculative investment fever of the 1920s. Among other things, they established elegant schemes by which control of vast utility empires could be maintained by ownership of only a few shares of stock at the highest level of the system – in essence a pyramid scheme. The best example of such an offending company was the Electric Bond and Share Company. As the country's largest, this holding company owned seventy-nine geographically scattered electric utilities as well as banking, insurance, and service organizations.[42] Many of these firms collapsed during the Great Depression, destroying for a time the public's faith in the private utility industry and providing renewed interest in the public power movement that led to the creation of the Tennessee Valley Authority in 1933.

The reorganization of the industry that resulted from the Public Utility Holding Company Act of 1935 abolished leveraged holding companies that had been established primarily to profit from inflated valuations of operating companies, watered stock, and speculative fever.[43] Nevertheless, the new law still recognized the benefits of economies of scale for large, interconnected systems. It therefore permitted the existence of holding companies that linked together several firms for providing power to contiguous geographical regions. The American Gas and Electric Company, formerly a part of Electric Bond and Share, continued to exist as an entity, holding nine operating companies located in adjacent areas. Renamed the American Electric Power Company in 1958, it still constitutes one of the largest electric utility holding companies in the United States, serving customers in the midwest and south central states.[44]

By the 1930s, then, the basic structure and strategy of the electric utility industry had been established. Like most other nascent industries, the electric utility industry evolved from a fluid state, in which entrepreneurs tried different approaches to gain mastery. Expanding upon the notion of a "dominant design" in the evolution of a technology, the utility industry developed a dominant design of technology, management principles, and organization. The technology of choice included turbine generators and alternating current. Even though water power still retained a foothold in power production, fossil-fuel burning had taken the largest share of the effort – 60% by 1929.[45] Meanwhile, business management used advancing technology and the grow-and-build strategy to create large, monopolistic power systems that had been legitimized by state regulatory bodies. And to raise the huge amounts of money needed for the capital-hungry industry, managers employed the financial and organizational structure of the holding company.

While this dominant design for the utility industry may appear obvious today, it was not at the time, nor was it inevitable. After all, it is conceivable that the industry could have developed along the lines that Edison had envisaged, with individual businesses generating electricity themselves in a decentralized fashion. Alternatively, the political climate could have made large-scale integration impossible, as was the case in England.[46] In the United States, too, the large regional system may have been excluded as an option if municipal systems had flourished, providing power only to small geographic regions bounded by city lines. Despite these possibilities and the simmering feud between public and private power systems, a specific dominant design did emerge in the United States, and it remained intact and largely unquestioned until the 1960s.

2
Establishment of a management culture

Beyond basic principles of operation, the early years of the twentieth century witnessed the development of a unique mind-set and behavior pattern for utility managers. Believing that electricity constituted a "magic" substance, transforming and improving the lives of everyone who used it, managers felt they had a special calling. As a result, they developed assumptions, historical lessons, and values that shaped the way they viewed their industry and problems that beset it. Altogether, these elements are often called "culture." While the term has recently become somewhat of a buzzword in the business literature, the notion of a culture within a corporate or industrial community is useful for understanding how people behave. A culture does not in itself cause events to occur, but when it fosters the creation of beliefs and generally accepted business practices, it is *one* contributing (though indirect) cause of action, and it helps explain why the utility industry managed itself differently from others. Firmly entrenched by the 1930s, an electric power culture became as much a foundation for managers as did the grow-and-build strategy and the steam-turbine generator.

Creation of an electric utility culture

The origins of the utility culture derive from the sense of mystery and awe associated with the phenomenon of electricity. At least from the days of the Enlightenment, electricity tempted the curiosity of popular audiences and royal courts. The 1700s witnessed a burst of activity in the study of the phenomenon, with Benjamin Franklin's experiments on static electricity bearing the most fruit. While developing a consistent theory for understanding the phenomenon, the Philadelphian also identified lightning as a form of electricity. In so doing, Franklin was hailed as the "tamer" of the power of Prometheus, who handed fire down from the heavens to be controlled by humankind. The mythical symbolism that accompanied Franklin's accomplishment may be considered overly dramatic, but it was just the first of many forms of symbolism that glorified the powers of electricity.

The discovery of current electricity in the twitch of a frog's leg by Luigi Galvani in 1800 increased the sense of wonder in the phenomenon. During the first half of the nineteenth century, scientists developed a body of experience and theory on the subject, while also introducing mechanical devices that took advantage of electricity. Batteries, telegraph systems, arc lights, electric tickers, and a myriad of other gizmos, some of which actually became successful commercial products, flooded the patent offices of the world. Among the most useful devices for the ultimate creation of the power industry was the dynamo, which produced electricity from mechanical motion, and the electric motor, which did the reverse.

Providing an alternative to the steam engine – the workhorse of the industrial revolution – the electric motor became the focus of numerous investigations in the early part of the century. In a burst of enthusiasm, the young James Prescott Joule, one of the many formulators of the conservation-of-energy principle during the late 1830s and 1840s, thought that he had discovered a way to produce an infinite amount of power from an electrical motor.[1] Such enthusiasm can be excused in a period when scientists and amateurs elicited the secrets of electricity. It is further reflected by the statement of Thomas Ewbank, Commissioner of Patents, in 1849: "The belief is a growing one that electricity in one or more of its manifestations is ordained to effect the mightiest of revolutions in human affairs."[2]

During the last two decades of the nineteenth century, the ideology of the electric utility industry became more fully established and prevalent. Much of the responsibility for the popularity of electrical science belongs to the "Wizard of Menlo Park," Thomas Edison. A colorful, hardworking, and informally trained man, Edison garnered the largest number of patents ever to be awarded one man in the United States. His legacy includes motion pictures, the phonograph, the storage battery, and the incandescent light bulb. Edison was also a confident and boisterous entrepreneur, declaring in 1879, for example, that he would create *de novo* an industry for the large-scale production of electricity even before he had invented the light bulb. His successful fulfillment of the promise in 1882, with the production of electricity at the Pearl Street, New York, station, marked the recognized beginning of the modern electric utility industry. The image of Edison as wizard remained a powerful force and incentive for newly minted electrical engineers of the late nineteenth and early twentieth centuries.

Almost from the start of the power industry, experts and laymen alike predicted the bounteous benefits that electricity would provide. Electricity was more than just a commodity, proclaimed the champions; rather it provided beneficial *services* to humankind in a way that no other product had done previously. Elihu Thomson, a cofounder of the Thomson-Houston company (which in 1892 absorbed Edison's company), adopted this theme as he outlined the future in an 1892 article for *New England Magazine*. After a philosophical discourse on how "true" scientific inquiry affords greater wisdom into the harmonies of nature while also bringing practical benefits, he foretold of widespread use of the telephone, incandescent light, and electric streetcar – all inventions that had begun to see usage. But he also detailed plans to use electricity for wireless communications, drilling in mines, metalworking and refining, home heating and cooking, and even in farming. Thomson found the last suggestion especially appealing. Recognizing a connection between electricity and chemical action, he thought plants would grow better by passing currents through the soil, yielding such delicacies as "pomme de terre à la dynamo" and "asperge electriques." In all his descriptions, Thomson offered a view of electricity as a powerful agent that beneficially served humanity.[3]

Continued demonstrations of electricity's practical uses spurred other writers to extoll the virtues of the "invisible fluid." In a popular article of 1901, "Electricity in the Service of Man," A. N. Brady explicitly tendered the theme of "service."[4] Like Thomson, Brady recounted the myriad of tasks that electricity had begun to perform, from hauling passengers on street railways (with trunk line electric railroads on the near horizon) to lifting people in elevators. In general, "electricity is coming closer home to the masses of the people, lifting their burdens, shortening and smoothing their journeys, lighting their streets and roads, and making their working tasks less difficult."[5] And all this in just twenty years – from the development of "a tiny laboratory spark into a powerful agent which moves the wheels of industry and commerce."[6]

For those not yet sophisticated enough to read the popular adult journals of the turn of the century, other ways existed to receive the message that electricity constituted a powerful force of nature that could revolutionize society.[7] Young boys growing up in this period could pore over L. Frank Baum's *The Master Key*, an "electrical fairy tale founded upon the mysteries of electricity and the optimism of its devotees."[8] Written in 1901 by the author of *The Wizard of Oz*, the story relates the experiences of Rob Jocelyn, an exuberant youth who toyed with batteries, wires, and motors in his attic. Though his mother objected to his playing with what appeared to be dangerous devices, Rob's father intervened, reminding her that electricity "is destined to become the motive power of the world. The future advance of civilization will be along electrical lines. Our boy may become a great inventor and astonish the world with his wonderful creations." Buttressed by his father's confidence, Rob accidentally conjured up the "Demon of Electricity" – really a benevolent genie – while connecting two live wires. Finally let out of the "bottle," the genie rewarded the boy with electrical devices that could sustain, maim, heal, protect, and transport. The story reveals that electricity can be used for good or evil purposes, but one can imagine that the book inspired experimental forays by youthful "Edisons" who tried to discover the right connection that would reveal the genie to them as well.[9] The book related the message that electricity is a slave to humankind, and that when used with wisdom, it can raise the standards of living for everyone.

In another form of literature, too, electricity and the devices it powered also became viewed in a positive manner. Utopian writers at the end of the nineteenth century portrayed future societies free of disease, misery, and even death through the use of electrical devices. In John MacMillan Brown's *Limanora: The Island of Progress*, electrically operated machines such as the "electrograph" read people's dreams, and the "historoscope" projected moving pictures of history on walls. More impressive still was the "alclirolan," a "radiographic cinematograph" combining a "microscope, camera in vacuo, and electric power" that revealed the secrets of microbes so they could be defeated as disease-causing agents.[10] Another utopian also envisioned electricity to be a central "force" in an imagined future society:

> [S]moke, cinders, and ashes are unknown because electricity is used now for all purposes for which formerly fires had to be built; our buildings and furniture, made of lacquered aluminum and glass, are cleansed by delicately constructed machinery that operates automatically. The very germs of unclean matter are removed by the most powerful disinfectants, electrified water, that is sprayed over our walls, and penetrates into every crack and crevice.[11]

In general, turn-of-the-century utopians used electricity in their literature to eradicate disease, to power efficient transportation and communications systems, and to make life easier through the use of home appliances.[12]

Symbolic representations of electricity early in the twentieth century also reflected the awe and respect accorded to the new form of energy. Romantic personages, such as the Goddess Electricity, graced the 1900 Paris International Exposition, extolling the majesty of electricity. Reigning in the Palace of Electricity, wrote historian Rosalind Williams, the goddess appeared poised in "a chariot pulled by Pegasus (representing the poetry of science) and a winged dragon (representing its material might), all of which was framed by a lacy metal-and-glass sunburst, while 180 feet below, at the base of the palace, a crashing waterfall made a liquid grand staircase for the Goddess."[13] Sixteen years later, another symbol of electricity's power became a landmark atop the American Telephone and Telegraph Company's building in New York City. Originally named the "Genius of Electricity" (and renamed "The Spirit of Communication" when installed in 1983 in AT&T's new building) the sixteen-ton bronze figure grips lightning bolts in one hand while electrical cables wrap his naked, winged body (Figure 9).[14]

The reality of electricity and the ideology of growth

Folklore and expectations converged with reality quickly. Unlike other forecasts of the future that littered the popular literature, many of these views of the foreseen benefits of electric power proved remarkably accurate. Electricity *did* become the motive power of commerce and industry.[15] In manufacturing processes alone, electrical motors moved into factories, replacing steam engines in basements and eventually obviating the need for belts, pulleys, and shafts as motors directly powered machines for producing goods. Energy consumption decreased, but more importantly, factories could be reorganized so that the process of manufacturing took precedence over the source of power generation. Increased flow of production, better working environments, improved machine control, and ease of plant expansion resulted directly from increased use of electric motors in factories. Plant managers and electrical engineers often reported productivity gains of 20 to 30% as early as 1901, "with the same floor space, machinery, and number of workmen. This is the most important advantage of all, because it secures an increase in income without any increase in investment, labor, or expense. . . ."[16] In short, electrification of manufacturing facilities increased labor, capital, and fuel productivity.[17]

Outside the factory, electricity also wrought a revolution. In the home, electrical appliances pumped the water, cleaned the carpets, preserved and cooked the food and, in general, helped to restructure family life. Electric utility companies encouraged the rapid growth of home electrification and power consumption by selling, renting, and lending a variety of new appliances.[18] As a result, the industry found that by 1933, domestic appliance operation accounted for 60% of residents' usage of electricity, up from 35% in 1926.[19] By 1941, and despite the effect of the Great Depression, home electricity usage continued to increase, with refrigerator use becoming especially popular. Since 1930, prices fell for the appliance as production facilities and scale economies improved. Technical improvements in the refrigerator also reduced its operating cost, "making it generally more economical to use a refrigerator than ice. . . ."[20] To urban residents at least, the material standard of living appeared to be improving as electricity service became prevalent in the home.[21]

On the farm, electricity brought dramatic change as well, though it took

Figure 9. "The Genius of Electricity," now residing in the main hall of AT&T's new corporate headquarters building in New York City. Reproduced with permission of American Telegraph and Telephone Corporate Archive.

longer in the making. Reluctant to invest in the expensive transmission and distribution facilities needed to serve the sparsely settled rural homesteads, private electric utilities (with some exceptions, notably Insull's Middle West Utilities Company[22]) powered only about 10% of the nation's farms in 1930. But with the Great Depression and the creation of the Tennessee Valley Authority and the Rural Electrification Administration, electric power began to make a headway in rural areas. Using electricity for their productivity-enhancing equipment as well as the "luxuries" of indoor life, farmers moved quickly to join their urban counterparts in the twentieth century. By 1945, almost six million farms – 75% of the total – received electrical service.[23] The rapidity of the transformation of farm life astonished observers. In a 1952 book, *The Farmer Takes a Hand*, the author commented on the "[a]stonishingly swift social revolution" that farm electrification had inaugurated. Moreover, he predicted, the "process of change is as yet only in the initial phase."[24]

Utility managers easily accepted these positive popular notions about the value of electricity for two reasons. First of all, many of them grew up during the period of electrification and increased appliance use in the early twentieth century, and they simply witnessed the remarkable changes wrought by use of the new commodity. In other words, they did not need to be convinced of the value of electricity. They knew it by experience. In fact, this experience may have constituted one reason why they wanted to work in the industry. Secondly, by the time they entered the employment of utility companies, the managers had already developed a professional culture that would have fostered similarly positive views about their new work anyway. Mostly trained as engineers, utility managers shared many of the general beliefs of the engineering community that had arisen during the period of professionalization at the end of the nineteenth and early twentieth centuries. These included the view that engineers made up an elite segment of society and that they possessed a legitimate technical approach that served as *the* way to solve social problems. Like Herbert Hoover, who became an exemplar of the successful, erudite engineer because he employed a rational and objective approach during World War I and as Secretary of Commerce in the 1920s, engineers saw themselves as stewards of technological and social progress who enhanced the public's welfare.[25]

So even as an ideology of electric power was forming, the profession that subsumed electric utility engineer-managers had already created a similar ideology. Engineers in general did "good" things for society, and utility engineers specifically produced a special commodity that enriched industrial and domestic life. The view is reflected by Samuel Insull, who could hardly have been accused of overstatement by his colleagues when he declared in 1914 that the young utility industry had already become "the greatest industry in public service today."[26] Reiterating the same notion 65 years later, Edwin Vennard – a utility manager since the 1920s – dedicated his book on utility management to "the raising of living standards everywhere."[27] In short, because their activities truly seemed to exemplify the best elements of engineering values, utility engineers clung to the popular views associated with the liberating power of electricity.

The ideology of electricity embraced more than just a realization of the commodity's usefulness. It included the corollary belief that people should use increasing amounts of electricity for a growing number of tasks. Of course, selfish reasons partially supported this view. Very simply, increasing usage meant a greater need for new generating equipment. As manufacturers competed for the power-machinery market, they developed larger and more efficient technology for converting fuel into electricity. These technological advances provided increased profits for companies and a gradual decline in the price of electricity for customers. In short, growth constituted an essential element of the grow-and-build strategy. Individually and collectively (through the National Electric Light Association – an industry trade organization), utilities pursued the goal of growth by engaging in a series of publicity and propaganda activities throughout the years.[28]

But the promotion of growth in usage went beyond being a good business strategy. It quickly became part of utility culture because it reflected the view that electricity usage meant better lives for customers. Moreover, growing consumption simply was part of the "American way of life."[29] President Truman's Materials Policy Cómmission accurately portrayed this common view in 1952. In its preface to a report on the future availabilities of resources and energy, the Commission noted that it shared "the belief of the American people in the principle of Growth [sic]. Granting that we cannot find any absolute reason for this belief, we admit that to our Western minds it seems preferable to any opposite, which to us implies stagnation and decay."[30]

The belief in growing electricity consumption also fit in well with other American values. As strongly materialistic people, Americans have associated "the good life" with acquisition of tangible products that provide comfort and cleanliness. Many of these devices, such as water heaters and other appliances, require electricity. Other common values, such as success, competitiveness, and external conformity encourage conspicuous consumption of products, many of which use electricity or are produced in factories that consume electricity in large quantities. Furthermore, the American sense of efficiency is also linked with increased electricity usage because of the discovery that electrically driven machines in factories increase productivity, thereby making manufacturing techniques more efficient. Finally, American culture places a high value on progress in the technical realm because of people's fundamental optimism about the future, with "progress" meaning "change" in a beneficial direction.[31] And more electrification meant more progress. Forty-six years after he created the first central power station, Thomas Edison still waxed eloquently over the benefits proffered by electricity. "So long as there remains a single task being done by men or women which electricity could do as well," he argued in 1928, "so long will that development [of further electrification] be incomplete. What this development will mean in comfort, leisure, and in the opportunity for the larger life of the spirit, we have only begun to realize."[32] In short, the promotion of electricity usage meshed well with the broader values of American society. It easily became part of the ideology of the electric utility managers.

Private power companies like Insull's Commonwealth Edison practiced what Edison preached by encouraging electricity usage among customers. But the Tennessee Valley Authority (TVA) provided the best and most extreme example of the beneficial effects of promoting electricity usage. A government agency created during the depth of the Great Depression, the TVA had a mandate to improve the lives of people who lived in a depressed region. Cheap electricity appeared to be one prescription for revival. When in 1933 the average cost of residential electricity in the United States stood at 5.5 cents/kWh, the promotional TVA rates *started* at 3 cents/kWh and declined to just 0.4 cents/kWh for usage over 400 kWh. From the beginning, a press release noted, rates had been "designed to encourage and make possible the widest use of electric service, with all the individual and community benefits which go with such wide use."[33]

The TVA approach to electricity promotion owed as much to ideological enthusiasm as to reason, but it appeared to prove a dramatic success nevertheless. Electric pumps provided indoor plumbing; electric stoves and heaters replaced dirty coal-burning ovens; refrigeration units extended the life of dairy products; and new industries descended upon the region, providing jobs to thousands.[34] "The people were quick in putting the low-cost electricity to work," recalled G. O. Wessenauer, TVA's manager of Power from 1945 to 1970. In rural Mississippi, he remembered, newly connected customers rushed to buy washing machines as their first appliance. They often put the machines, which served as status symbols, on their front porch so that passersby would know they had TVA electricity.[35]

As increased electricity usage correlated with the Valley's rejuvenation, and as the rural population began enjoying a standard of living that previously only could be attained by city dwellers, the TVA experience offered another powerful example for the value of electricity and a new rationale for encouraging electricity usage. Within the TVA, promotion of load growth achieved an almost religious and evangelical status. As noted in 1944 by TVA chairman David E. Lilienthal, the incredibly large amount of power used by the region's customers – 50% greater than the nation's per capita average – has

> deep human importance, for this must be remembered: the quantity of electrical energy in the hands of people is a modern measure of the people's command over their resources and the best single measure of their productiveness, their opportunities for industrialization, their potentialities for the future. A kilowatt hour of electricity is a modern slave, working tirelessly for men. Each kilowatt hour is estimated to be the equivalent of ten hours of human energy; the valley's twelve

billion kilowatt hours can be thought of as 120 billion man hours applied to the resources of a single region! This is the way by which, in the Age of Electricity, human energies are multiplied.[36]

Outside the agency, the TVA experience of promotion reinforced and extended an important lesson earlier taught by Insull. While private utility managers believed in the value of electricity and its promotion, few realized that electricity demand had such a high elasticity. In other words, TVA planners demonstrated that usage could be stimulated dramatically with extremely low rates, which would be justified only if usage actually increased through promotion and appliance saturation.[37] Private utilities might never want to set rates as low as TVA's, but the agency's experience showed that lower rates brought increased consumption and revenues to compensate for the lower per-kilowatt-hour prices. During the Depression, this lesson suggested that rate cuts would not necessarily mean decreased income and financial instability. After the Depression and World War II, the lesson implied that promotional rates could be used to pursue the grow-and-build strategy with even more gusto.[38]

By the 1930s, then, the culture of utility managers had become well established, and it provided the third leg for the industry's foundation along with basic management principles and large-scale technology. Of course, all three legs were interrelated: business strategy called for creation of regulated monopolies serving a diverse set of customers who used increasing amounts of electricity. And huge generation technology providing economies of scale offered the means to supply the demand and increase productivity. Finally, the culture and ideology emerged to justify and reinforce management principles and the use of large-scale technology. Managers could now think positively and proudly about their industry and its contributions to society. As agents of technological and social progress, utility managers knew they provided for the country's economic well-being, and they could feel that all their activities fit into a coherent and consistent framework. For another reason, too, the culture proved extremely important: even when some of the management principles and technology came into question in the 1970s, the culture – and the assumptions about growth that it subsumed – would remain intact.

3 Manufacturers and technological progress before World War II

A small group of equipment manufacturers designed and built the technology that became an integral part of the utilities' grow-and-build strategy. Responding to and anticipating the needs of their customers, manufacturers made progress along many fronts simultaneously, with improvements produced incrementally in boilers, turbines, generators, and transmission and distribution systems. Generating costs contributed about 80% of the total costs of producing power,[1] and most of the gains resulted from improvements in just two areas: obtaining economies of scale from more powerful generating units, and raising thermal efficiencies so that less fuel was needed to produce each kilowatt-hour of electricity.[2] Meanwhile, advances in transmission and interconnection technologies enabled innovations in generation to be exploited by individual utility companies and regional associations of firms.

During the first fifty years of the twentieth century, manufacturers designed technology in what can be considered a traditional manner for engineers: they manufactured new versions of technology and then allowed customers to try them "in the field" before moving on to the next advance. The motivation for such a conservative approach came largely from the utilities, which demanded reliable and well-tested equipment that would provide long-lasting value for their huge capital investments. The manufacturing style can be called "design by experience," and it paid off handsomely in small gains that accumulated over extended periods of time.

Manufacturing environment and design by experience

The rapid expansion of the electric utility industry in the early 1900s created the environment in which manufacturers could develop new technology. Emulating Insull's success with the grow-and-build strategy, utilities garnered new customers and sold them increasing amounts of electricity. Between 1900 and 1920, the utility industry boosted its capacity twelvefold, increased its energy sales sixteenfold, and escalated its capital investment from $500 million to $4.5 billion.[3] Rapid expansion continued during the booming 1920s, which saw utility capacity grow from 12,000 to 30,000 MW and usage more than double.[4]

During this period of growth – and for several decades thereafter – utility managers desired three features from new technology. First, they wanted reliability. Boilers, turbines, generators, and auxiliary equipment needed to operate as promised and not break down unexpectedly. Unreliable equipment cost utilities severely by requiring excessive maintenance and construction of other facilities for backup reserve. Secondly, they wanted machines that offered greater power at lower unit costs. In other words, they depended on economies of scale to mitigate growing expenses for construction and labor. Finally, they required higher thermal efficiency from their plants, so that they could save on fuel costs by producing more electricity with the same amount of raw energy.[5] These last two requirements proved essential if utilities hoped to pursue the grow-and-build strategy.

Manufacturers obviously had to respond to these market demands if they wanted to remain in business. In general, they succeeded admirably by performing most of the research and development work on new technology. (In contrast, utilities performed work on their systems, which integrated the new technology into the power network.) Among the manufacturers – also known as suppliers or vendors – two companies stood out in the turbine-generator business: the General Electric Company and the Westinghouse Electric Corporation. Both firms began producing equipment for the fledgling electric power business as soon as it began in the late nineteenth century, and they competed for market share by (among other things) offering equipment to utility companies that advanced the state of the art. Westinghouse took an early lead by becoming the first in 1898 to offer small steam turbines to utilities, based on the Parsons design, followed by units as large as 3.5 MW in 1901. Spurred by the competition, General Electric pursued development in 1897 of a turbine designed by the American, Charles Curtis, delivering its first 1.5-MW unit in 1903. Launching an aggressive sales campaign in 1903, GE grabbed

80% of the market share in one step. Settling at about 65% for the years before 1925, GE's market share remained remarkably consistent throughout the twentieth century, with Westinghouse resting in second place with about 25% of the share. As first movers in an industry that proved to be heavily capital-intensive, these firms had tremendous advantages over newcomers. Nevertheless, the companies encountered some competition from Allis-Chalmers, a diversified equipment manufacturer founded in 1901, which gained about 9% of the market share.[6] A few foreign firms, such as the British Thomson-Houston company and the American affiliate of the Swiss company, Brown Boveri, picked up the remaining 1% of the market. Meanwhile, boiler technology advanced largely through the efforts of two other companies, Combustion Engineering and Babcock and Wilcox.[7]

As a result of the demanding market created by utilities, manufacturers developed their new machines using a "design-by-experience" technique. Vendors would introduce innovations in a pioneering technology that was custom-made for a utility having unusual requirements. The design process took a few years, as did manufacturing. The machine would then be put into service and observed by the utility and manufacturer. Based on experience with the equipment, the vendor designed another version of an incrementally better one for other customers. As one engineering manager recently put it, the design philosophy meant "making the next one like the last one that worked" – a reasonable approach that moved from one success to another.[8] Over the long run, the advances appeared large, but the manufacturers took modest incremental steps slowly enough so that they could develop experience and so that users could gain confidence in the new design. As a GE engineer pointed out in 1930:

> It is worth while to notice that the rapid progress which has been achieved in this country has been made in steps that have not been so large as to go beyond the possibility of producing reliable equipment. Each step, although large from some viewpoints, has not gone so much beyond experience as to lead to the incorporation of many untried elements in the vital parts of the construction. The result has been that all stations of new and untried designs have been successful commercially, and are running in regular commercial service today. . . . The art is still progressing very rapidly, and the indications are that it will continue to do so indefinitely.[9]

Other similar pronouncements abound in the literature. Shunning too-rapid acceleration of new designs, for example, one *Electrical World* editor in 1929 decried efforts made by a German engineer to boost the pressure of steam for turbines to greater than 2,000 pounds per square inch (psi) at a time when American engineers employed steam at little more than 1,200 psi. While the increased steam pressure would lead to greater thermal efficiency of the plant *in theory*, little experience existed yet, and the jump appeared too radical. "Experimentation should go on with high pressures and with other possibilities," noted the author, "but major power stations should be built in the light of cost and operating data available in existing plants. There is no need for opinion when facts are available."[10] And in discussing specific metallurgical problems that might inhibit use of high-temperature steam, another industry insider concluded that a gradual approach to progress seemed most acceptable: "It is not to be doubted that the answers to these many questions [concerning metallurgy] will lead to high-temperature practices, but engineers are not minded to experiment much without knowing the ground they are treading."[11] The reluctance to accept wholeheartedly the most novel designs without first acquiring experience is further demonstrated in descriptions of design-parameter limits, such as steam pressure, that exceeded "conservative good practice."[12] Finally, as one GE engineer discovered, when faced with choices for new technologies, many utility managers will "desire to avoid, for obvious reasons, the introduction of new and not yet completely proved types of equipment."[13]

This design-by-experience approach proved necessary, especially in the century's early decades, because sophisticated design theory simply did not exist. As with many other engineering endeavors, the people who built these complex machines learned as they went along. They built equipment and tested it in their shops or in the field, learning about such things as practical limits to vertically mounted steam turbines. (General Electric ceased producing these machines after shipping a 20-MW unit in 1913.[14]) And because they could not understand everything about their hardware and the environment in which it operated, engineers typically compensated by building in extra margins for error. The practice led to machines that the literature described as having "generous" designs. As noted by Alvin Weinberg, a leading figure in the development of nuclear energy, engineers typically use conservative techniques when building new or large-scale technologies because they "cannot afford the luxury of examining every question to the degree which scientific rigor would demand. Indeed, 'engineering judgement' connotes this ability, as well as necessity, to come to good decisions with whatever scientific data are at hand."[15] In the power-equipment manufacturing business, as in many others, engineers commonly overdesigned equipment to ensure reliability. As an example, a Westinghouse turbine sold in 1900 had been rated at 1,500 kW. But because of its overdesign, the customer utility found that it could use the machine at a 2,000-kW rating – or 33% greater than originally anticipated.[16] The over-design practice would remain common for several decades. As will be seen later, this practice – which formed part of the design-by-experience approach – would be abandoned after World War II.

Learning occurred quickly during periods of rapid growth, when utility companies upgraded their technology base frequently to meet increasing demand. In 1929 and 1930, for example, utilities installed more than forty fossil-fueled units each year, compared to less than ten each year a decade earlier.[17] As a result, manufacturers enjoyed the opportunity of offering a large variety of new advances as utilities made recurrent purchases, with each new piece of equipment sporting a novel feature. In other words, the rapid growth rate meant that utilities could assimilate improvements more quickly than during periods of slow expansion. With capacity needs doubling in six years, one observer noted in 1930 that "experience tells us with considerable certainty that obsolescence will take place faster in the next six years than was the case in the preceding twelve years, just as the preceding twelve years is comparable in this respect to the still earlier twenty-four years. We must recognize the extent to which human activities are speeding up."[18] Despite the expectation of quick obsolescence, utilities still prized long-lived technology that had demonstrated its dependability in the field. Because of this requirement, utilities preferred new technology that incorporated small leaps in performance.

Not all utilities, of course, could take advantage of the most advanced and largest technology. As noted in 1930, while the size of the largest turbine unit had increased dramatically in the previous two decades, the average size of units sold in the United States increased much more slowly. This occurred because firms serving small communities that were not yet interconnected into regional systems simply did not require the largest equipment available. Nevertheless, the less powerful unit incorporated many of the improvements of the pioneering technology. In designing large-capacity transformers, for example, manufacturers employed a "top down" approach in which "changes and improvements are apt to be made first in larger transformers and, after thoroughly tried out in these large transformers, are later extended to small transformers."[19]

In developing technology "by experience," manufacturers obviously needed the assistance of their utility customers. As such, utilities played a role in spurring technological advances by serving as partners in risk taking with new equipment. Suppliers were naturally anxious to install prototype generating units in order to gain experience that would help them create more advanced designs. Utilities, on the other hand, wanted to benefit from new technology. As will be noted later, several utilities willingly joined into risk-sharing relationships with manufacturers. In the early years of the century, the relationship resulted in part from the ownership by manufacturing companies of utility securities. The Electric Bond and Share Company, a wholly owned holding company subsidiary of General Electric from 1905 to 1924, possessed 12.5% of the installed generating capacity in the United States by the mid-1920s. As can be imagined, the Electric Bond and Share Company's operating firms bought equipment largely from GE and worked together with the manufacturer.[20] The vendors therefore enjoyed an environment in which experience and learning could accumulate in tandem.

Technical advances in scale frontiers

Insull apparently understood the benefits of increasing the scale of his power technology. First of all, he knew that simple geometrical expansion played an important role in providing good economic returns when dealing with power technology. By doubling the diameter (and circumference) of a steam delivery pipe, for example, the volume of the vapor that can be contained is quadrupled. The same principle applies when constructing several other components of a power unit: with modest increases in material inputs, disproportionately large outputs are obtained. One rule of thumb – the "six-tenths" rule – reflected this geometrical fact by approximating the added capital cost of capacity at 0.6% for each 1% of capacity increase.[21] Beyond scale economies derived from geometry, however, are others that come from simplified organization. Building one large unit instead of two smaller ones requires only one foundation for a turbine, one control room, and less auxiliary equipment, yielding lower capital costs per unit of capacity. In addition, utilities benefited from scale economies resulting from the reduced number of operating and maintenance personnel required to run a larger-capacity unit. Since the operation of a large unit from a single control room did not differ significantly from operation of a smaller unit, for example, the company saved by having staff for only one control room. Together, these capital and operating economies of scale proved attractive to many companies, such as a Milwaukee utility in 1928, which purchased one 60-MW unit rather than two 30-MW machines.[22] As a General Electric engineer reported in 1930, studies "made by many groups of engineers and under many conditions of design and location, indicate conclusively that the overall first cost of a station per kilowatt of installed capacity decreases markedly with an increase in the capacity of the turbine-generator sets, and with an increase in the capacity of the boilers used."[23] For this reason, utility managers desired large-scale technology.[24]

Turbines constituted the core of generating technology, converting steam flow into rotary motion for use by generators. The earliest steam turbines looked much like their water turbine counterparts, with their shaft aligned vertically and driving a generator below it. But problems with balancing the apparatus suggested that enlargement of the turbine would require a horizontal position for the turbine and generator. After being introduced in 1910, the horizontal turbine overtook the size of vertical machines by 1913 when they "grew" larger than 20 MW. Utilities purchased horizontal steam turbines exclusively after 1914.[25]

For utilities that wanted to obtain more than about 80 MW from a turbine, however, problems existed with *generators*. Throughout the 1920s, these machines simply could not be made to produce more than about 100,000 kilovolt-amperes (equivalent to 100,000 kilowatts) of electricity. As the machine that turned mechanical motion into electric current, the generator contained a heavy rotor known as the "field." Wrapped with wire and "excited" by direct current to produce a magnetic field, the shaft rotated within a "stator" – stationary bars of copper where current is produced. As the output of turbines increased, generator design became difficult because simple scaling up of components created overheating problems as the generators produced more current per unit volume of metal. The key to the problem's solution came by cooling the generators more efficiently. Standard cooling techniques simply took air at room temperature and passed it through the generator, absorbing heat as it passed over the stator bars and field windings. Until 1937, all generators used this cooling technique. But air has a relatively low heat capacity, meaning that a given volume absorbs a limited amount of heat. Four times as effective for cooling, hydrogen gas found use (despite its flammability) in a few machines before World War II.

By removing heat more effectively, hydrogen made possible the use of turbine generators operating at two or more times the speed of traditional units (1,200 to 1,800 revolutions per minute [rpm]). Three benefits accrued from higher speeds. First, turbines reached closer to their theoretical output as their design speed increased. Thus, in most cases, the thermodynamic efficiency of these turbines improved. More importantly, however, designers could engineer faster turbines with smaller blades, shells, and rotors, especially for high-pressure and -temperature machines. The smaller machines took up less space and weighed less than slower machines of the same capacity, thus reducing cost. While a 1,200-rpm turbine rotor used for producing 50 MW weighed almost 108,000 pounds, a 3,600-rpm rotor weighed only 35,000 pounds and produced the same amount of power.[26] Finally, fast turbines could be manufactured more easily and with greater reliability because deflections and distortions in materials decreased in smaller components. Already in 1940, the weight of evidence favored faster machines in terms of reliability.[27]

Despite these advances in fast turbines, the most powerful turbine generators still could not compare to the output of the slower machines. In 1930, for example, the largest-capacity 3,600-rpm machine yielded only about 15 MW.[28] For this reason, and because hydrogen cooling techniques had not yet become well accepted, utilities purchased units that did not consist of a single-shaft turbine and generator when they needed huge amounts of power. Instead, they bought units such as GE's gigantic (for the time) 208-MW machine built in 1929, which consisted of three sets of turbines and generators with steam passing from one to the next.[29] All components operated at 1,800 rpm.[30] Known as "cross-compound" units, they later were combined with faster units so that steam first passed through a turbine-generator system that rotated at 3,600 rpm and then through another rotating at 1,800 rpm. In both these cases, engineers developed designs that extracted the most energy from steam at the cost of using multiple generators and turbines.

Though larger than simpler single-shaft machines, cross-compound units still could yield great economies. In 1928, for example, the United Electric Light and Power Company in New York installed a complex cross-compound 160-MW unit built by the American Brown Boveri Company. Compared to a single-cylinder and -shaft design of 25 years earlier, which weighed 32 tons for each megawatt it produced, the new unit produced its power using only 8.8 tons per megawatt.[31] And it filled the same floor space as a previously used 30-MW machine.[32]

Boiler designers always had to play a game of "catch-up" with engineers who planned other components of power units. Boilers simply could not be built large enough to produce the volume of steam needed by increasingly large turbines and generators. It required eight Babcock and Wilcox boilers, for example, to supply steam for each of the 5,000-kW turbines that Samuel Insull installed in his Fisk Street Station in 1903.[33] A few years later, big plants that housed five 10,000-kW turbines commonly required fifty or sixty boilers, with all the attendant duplication of components and laborers.[34]

But the boiler manufacturers responded to this challenge. They introduced innovations such as the use of pulverized coal to achieve higher temperatures, for example, so that each boiler provided more heat output. Employing water-cooled furnace walls and new firebrick that resisted erosion, manufacturers found they could enlarge the output of boilers and reduce the number needed to produce steam for a turbine generator.[35] By the 1930s, manufacturers succeeded in designing single boilers that matched individual turbine-generator sets. In fact, "unit-type" construction – one boiler per turbine generator – became commonplace even in units as large as 125-MW capacity because of their high level of reliability.[36]

In general, utilities took advantage of manufacturers' abilities to increase the capacity of components. The economic benefits simply overwhelmed managers, because scaling up units yielded decreasing investment costs. In the fifteen years before 1929, for example, capital costs hovered around $100 per kilowatt as average unit capacity increased from 12.5 to 24 MW. Considering how the consumer price index rose during this period, the real cost of investment actually declined about 70%![37] No wonder utility managers desired large-scale technology.

Improvements in *transmission* systems also occurred stepwise throughout the power industry's first several decades. While comprising a relatively small portion of a power system's total capital cost and incurring low operating expenses, transmission systems nevertheless contributed significantly to providing lower cost and increasingly reliable service. They did this by operating at ever-increasing voltages and by permitting interconnection between different power plants owned by contiguous utility companies.

Transmission systems that emerged in the late nineteenth and early twentieth centuries functioned as point-to-point lines, providing power from distant hydroelectric sites to cities or industrial sites where the electricity could be used. Taking advantage of the principle that alternating current can be transmitted with fewer energy losses at high voltage, manufacturers such as Westinghouse created a system that sent electricity at 11,000 volts from Niagara Falls to Buffalo, New York, in 1896.[38] Through efforts made by General Electric and Westinghouse, transmission voltages increased, reaching about 60,000 volts in 1900, 150,000 volts in 1912, and 240,000 volts in 1930.[39] Increased voltages did more than reduce energy losses. In the same way that a large volume of water flows through a pipe when forced under high pressure, more electricity can pass through a transmission wire at high voltage than at low voltage. In recently designed systems, for example, a 115,000-volt system has the capability to conduct 60,000 volt-amperes of electricity, but a 230,000-volt wire can transmit 250,000 volt-amperes.[40] While striving for greater voltages, manufacturers encountered and dealt with several problems. To prevent "flashover" or short circuits between wires and transmission towers, for example, manufacturers needed to improve insulation techniques. Likewise, they also had to develop lightning arresters, transformers, circuit breakers, and switchgear that could withstand high voltages and changing environmental conditions.[41]

More important for the development of the power industry than point-to-point service, however, was the use of transmission lines to interconnect different utilities. Receiving a big boost during World War I, when some utilities connected their systems in order to avoid local power shortages, interconnection through high-voltage transmission played a crucial role in effecting economies in electricity supply. As Insull and others discovered, interconnection decreased the need for isolated systems to maintain high reserves of generating capacity. It also reduced generating requirements for individual companies due to the increased diversity of load, and it provided more reliable service in case of one plant's failure in the interconnected system.[42] Additionally, interconnected systems allowed for the concentration of large amounts of power in fewer plants so that economies of scale could be effected. By building a few large plants that offered low unit costs rather than operating several small, isolated plants, Insull already had learned that transmission facilities turned scale economies on the plant level into scale economies on the system level.[43] In short, the development of high-voltage transmission systems contributed as much to the steady increase in capacity of power production units as did advances in turbine speed or generator-cooling techniques.[44]

Technical advances in thermal-efficiency improvements

Scale economies provided dramatic productivity increases, but so did improvements in thermal efficiency. Here, too, advances came in incremental steps. But unlike scale economies, which were measured in financial terms, such as dollars per kilowatt capacity, thermal efficiency could be easily calculated in more "technical" engineering terms that derived from operational performance. Several measures existed and were used interchangeably: simple "efficiency" is defined as the amount of electrical energy that could be generated as a percentage of the total fuel energy. Alternatively, the "heat rate" equals the amount of heat, measured in British thermal units (Btus), needed to produce 1 kWh of electricity. Finally, the "coal rate" relates the number of pounds of coal necessary for yielding 1 kWh of electricity.[45] As all these measures showed improvement through the years, engineers derived a special pleasure in knowing that they contributed to progress in their industry.

Thermodynamic efficiency of a steam turbine-generator system is governed by the "Rankine cycle," in which burning fuel heats water, turning it into steam at high temperature and pressure. Steam expands through the turbine, and after converting its kinetic energy into rotary motion, it exhausts to a low-temperature condenser. Efficiency can be raised by using steam at the highest possible temperature and pressure and by lowering the condenser temperature. In practice, condenser temperature cannot be easily controlled, as it depends on the temperature of water from a river, lake, or other body of water. The inlet steam temperature and pressure, however, can be controlled by boilers that burn fuel most efficiently, and by turbines whose construction materials withstand extreme steam conditions.

Pressure and temperature increases of steam marked major advances in *boiler* design before World War II. When the Babcock and Wilcox company supplied boilers for Insull's first turbines in 1903, they operated at a pressure of 180 pounds per square inch and at a temperature of 530 °F.[46] As central station electricity generation caught on, however, utilities demanded boilers that burned fuel more efficiently. One significant innovation that pushed up temperatures consisted of pulverizing coal for combustion in water-cooled furnaces. In contrast to the previous method of burning chunks of coal on grates, the new method of grinding coal into fine powder before injection into the furnace increased the surface area of the fuel significantly, thereby aiding in combustion. The innovation in itself almost single handedly enabled boiler temperatures to increase from about 650 °F – where a plateau existed since 1915 – to 750 °F by the mid-1920s.[47]

The higher temperatures of boilers forced the next advance in boiler design. By-products of pulverized coal, especially fused ash, eroded the brickwork inside the furnace. At the same time, higher temperatures weakened the bricks and caused concern over the structural integrity of large boilers. One successful approach for resolving these problems consisted of cooling furnace walls with water. In such a furnace, water-carrying tubes that made up part of the steam-making system extended downward along the walls of the furnace into the zone of active combustion. The system performed two important functions: water cooled the furnace walls, allowing for higher combustion temperatures; and it heated water to be turned into steam. By the end of the 1920s, these advances, coupled with others, permitted the production of steam at 750 °F and up to 1,400 psi.[48]

High temperatures provided better efficiency, but they also added to metallurgical problems. While steels made with carbon worked well up to about 825 °F, they deformed under constant stress at higher temperatures in a process known as "creep." After several combinations of alloying agents were investigated, molybdenum proved effective in resisting creep. The late 1930s therefore saw temperatures reach just above 900 °F.[49]

After *turbines* became the prime mover of choice by about 1910, the major manufacturers undertook systematic efforts to understand turbine theory and practice. The American Society of Mechanical Engineers also established committees for investigating properties of steam, while manufacturers and research organizations studied the aerodynamics and vibration patterns of turbine blades. Simple improvements in nozzle shapes, understanding of surface finishes, and other detail improvements led to incremental advances of 10% or more in internal efficiency.[50] Higher pressures and temperatures also produced efficiency increases, leading to reduced usage of fuel for producing electricity.

By the onset of World War II, manufacturers had made tremendous progress in developing technology for the electric utility industry. Most of these advances occurred when rapid growth rates drove the industry before the Great Depression, and utilities took advantage of their monopoly status to employ ever-larger and more efficient generating technology. Power units "grew" from the unusual 5-MW unit operated by Insull in 1903 to the 208-MW cross-compound unit of 1929 – the largest single unit (though not a single-shaft machine) built until 1953.[51] The greatly increased capacity of units permitted fewer machines to do a larger amount of work. Using 20% fewer power stations in 1928 than in 1922, *Electrical World* proudly pointed out, the industry doubled its power-making capacity.[52] With the Depression creating excess capacity in the utility industry, however, demand for new power units dried up in the 1930s. This explains the plateauing at 160 MW for the most powerful single-shaft turbine generator after 1929. Not until the post-World War II period would another spurt in unit scale occur.

But while manufacturers did not continue to scale up power units, they continued their efforts to improve thermal efficiency. By taking advantage of stronger metals, boilers and turbines could withstand steam at higher temperatures and pressures, yielding tremendous improvements in thermal efficiency. From the days of Edison's Pearl Street Station (1882) and through the Depression, the average thermal efficiency of plants rose from about 2.5 to about 20% in 1940.[53] Overall, these design advances, channeled within a utility industry that depended on improving technology and increasing sales, brought down the cost and price of electricity. In nominal terms, the average resident paid about 22 cents in 1892 for a kilowatt-hour; in 1932, that same unit cost 5.6 cents – 75% less. Adjusted for general cost of living increases, the price declined 86%.[54] To industry insiders and outside observers, the systematic and incremental approach used in attaining improvements in power technology must have appeared highly impressive.

4

Postwar strategies of utilities and manufacturers

The exuberance that utility managers felt during the first three decades of the century moderated during the Great Depression. With slackening demand, financial insecurity, and widespread criticism of the industry resulting from holding company abuses, the primary activity of managers consisted of the less exciting, but necessary, task of avoiding bankruptcy and public takeovers. World War II furnished a new challenge, and utilities scrambled to produce power for a reinvigorated economy despite the limited amount of new equipment available. The early 1950s, however, witnessed a resurgence of interest in promoting electricity usage and in producing low-cost electricity through the use of large-scale technology. In other words, utility managers actively pursued the grow-and-build strategy again.

As utilities resumed their traditional strategy, manufacturers felt the need to produce new, larger-scale, and more efficient technology. Though utilities returned to their basic principles, the accelerated growth after World War II caused the manufacturers to modify theirs. Instead of the traditional design-by-experience approach that led to gradual improvements, the vendors developed a new strategy. Sometimes called "design by extrapolation," the approach enabled manufacturers to produce larger technologies more rapidly, contributing to a second spurt of scale increases after the war.

Utilities' promotion of electricity usage

As Insull discovered in the early 1900s, promotional rates offered to industrial customers facilitated the successful pursuit of the grow-and-build strategy. But beginning in the 1920s, as the domestic electrical appliance industry burgeoned, managers discovered that similar rate structures could be offered to residential users as well. Using more appliances throughout the day, residential customers helped utilities achieve better load factors and greater profits.[1] If they needed further motivation to promote electricity usage, managers could look to the TVA experience in the 1930s, which demonstrated that promotion of cheap power and appliance saturation could benefit customers and utilities alike.

The coming of World War II interrupted the concerted effort by most investor-owned utilities to increase electricity usage. Expanding defense industries and a reinvigorated economy rapidly drove up consumption without the need to promote it (after an increase of only 38% for the ten years after the Depression began[2]), and utilities did their best to provide adequate service. But even after the conflict, it took a few years before utilities regained the momentum of promoting electricity usage. For one thing, the industry experienced a shortage of capacity. Editorials in trade journals lamented the capacity crunch as usage jumped 14% from 1946 to 1947 and as reserve margins declined to an average 8%.[3] For the years 1947 and 1948, the margin slipped further to 5%, a dangerously low level at a time when managers considered 20% a safe margin.[4] "Instead of leaving a large capacity margin, disposal of which would require [an] extraordinary selling effort," noted *Electrical World* in 1949, "the postwar load came close to requiring the entire available capacity."[5]

The shortage occurred simply because the manufacturers that normally built steam turbines for power plants had diverted their efforts to produce marine turbines for warships.[6] Labor troubles among construction workers in 1946 and 1947 added to problems by making it difficult to install the equipment that manufacturers finally began producing again.[7] Utilities attempted to meet demand while waiting for new plants by installing quickly built internal combustion units. Only 45 such units (with an average output of 737 kW each) were used in 1945, but the number rose quickly to a peak of 194 units (with an average output of 1.1 MW each) in 1948.[8]

Once promotional activities began again, they did so cautiously. While many utility managers believed that electricity usage meant industrial growth and prosperity, they appear to have been most concerned about the financial health of their companies, some of which had just been reorganized as independent firms.[9] As one utility president explained, his company had embarked upon a $52 million construction program between 1946 and 1951, and he first wanted to make sure that electricity usage would be sufficient to make profitable use of the equipment. Aside from promoting electric lighting, the traditional use for electricity, the company instituted a "Go All-Electric" program to encourage electric refrigeration and electric cooking.[10] Selective in its promotion, the utility aimed at increasing the types of usage that would improve load factors and increase earnings.[11] As argued by Edwin Vennard, vice president of one of Insull's former companies, the Middle West Service Company, selective sales should be advocated "with emphasis not on gross kilowatt-hour sales, nor gross revenue, nor even net revenue, but upon percent return."[12]

Less selective promotional activities began again in earnest in the 1950s as managers realized that capacity additions improved reserve margins and as the industry returned to normal operations.[13] But they also were encouraged by the fervent zeal of the 1950s conjured up by the Korean War and anticommunism crusades. President Truman's Materials Policy Commission, convened in 1952 to evaluate the future availability of natural resources and energy supplies, portrayed accurately the sense of mission that motivated many Americans at the time:

> The United States, once criticized as the creator of a crassly materialistic order of things, is today throwing its might into the task of keeping alive the spirit of Man and helping beat back from the frontiers of the free world everywhere the threats of force and of a new Dark Age which rise from the Communist nations. In defeating this barbarian violence moral values will count most, but they must be supported by an ample materials base. Indeed, the interdependence of moral and material values has never been so completely demonstrated as today, when all the world has seen the narrowness of its escape from the now dead Nazi tyranny and has yet to know the breadth by which it will escape the live Communist one – both materialistic threats aimed to destroy moral and spiritual man.[14]

Reinforcing managers' belief in the value of electricity, this "call to action" further motivated companies to promote electricity usage as a means to provide for national security needs. Constituting one of the "resources for freedom," according to the report, electricity would power factories, energize homes, and supply the needs of the newly created Atomic Energy Commission. To do this, company managers felt they needed to sell aggressively and to make expansion of facilities a routine function.[15] Promoting electricity usage (and electrical appliances) meant that homes, industrial plants, and commercial buildings would be modernized and more efficient, providing "one sound reason why our American economy will never stop growing."[16] As noted later by the American Society of Mechanical Engineers, which awarded its prestigious Fritz medal in 1956 to Philip Sporn, president of the American Gas and Electric Company, "America was a dynamic society whose goals of human welfare were constantly rising. . . . Electric energy, to increase productivity and thus meet the demand for higher living standards, must be provided at a faster rate than that of population growth and manpower."[17] In short, a growing utility industry, supplying more electricity to the American public, provided the basis for a strong economy and a society. The view meshed well with cultural values already held by utility managers.

As residential users began to adopt a wider diversity of appliances than was previously believed, utilities sought ways to improve load factors for this group of customers that previously demonstrated the worst usage patterns. Televisions, for example, had already been installed in one million homes by mid-1949, and analysts predicted another 15 million sets in use within five years, adding 250 kWh of consumption per receiver at a time when average residential use remained under 1,700 kWh per year.[18] Room air conditioners also caught on quickly, jumping almost fortyfold from 43,000 sales in 1947 to 1,673,000 in 1958.[19] But promotion of these last appliances had one pitfall: they caused high peaks in usage during the summer with no compensating peaks at other times of the year, thereby reducing load factors still more. The solution, according to Philip Sporn, consisted of wider use of heat pumps or other forms of electrical heating in homes.[20] Though viewed with uncertainty just a few years earlier – studies in 1953 suggested that just 2% of homes in the 1960s would be electrically heated[21] – the effectiveness of electrical heating for increasing load factors became recognized through the mid-1950s.[22] The all-electric home – one that used no competing fuels – loomed large on the horizon.

Shifting the Emphasis from
Machines to People.

Figure 10. Reddy Kilowatt, the marketing character and corporate spokesman adopted by scores of investor-owned utility companies. Illustration provided by Reddy Communications, Inc. Reddy Kilowatt® is a federally registered trademark and service mark and is used here with the permission of Reddy Communications, Inc.

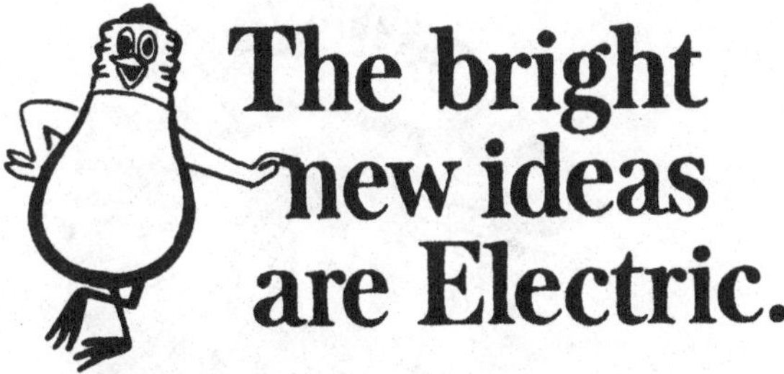
The bright
new ideas
are Electric.

Figure 11. "Little Bill," the marketing trademark used by Chicago's Commonwealth Edison Company in the 1950s and 1960s. Little Bill® and the slogan, "The Bright New Ideas Are Electric,"® are valid registered trademarks of the Commonwealth Edison Company and are used with permission from Commonwealth Edison.

To help push the message of the benefits of greater electricity consumption, some utilities increased their use of the spry cartoon character "Reddy Kilowatt" (Figure 10). Created and copyrighted in 1926 by Ashton B. Collins, then commercial manager of the Alabama Power Company, Reddy became a ubiquitous marketing tool, mascot, and spokesman for those investor-owned utilities willing to pay licensing fees. Within nine months of his introduction in 1934, Reddy had already been signed up by several of the country's larger utilities, including Philadelphia Electric, Pennsylvania Power and Light, Ohio Edison, and Tennessee Electric Power.[23] By the 1940s, more than 200 companies used him to promote electricity usage and deliver other corporate messages.[24] According to a history written by the symbol's owner, Reddy Kilowatt (renamed Reddy Communications), Inc., Reddy "dramatizes the benefits of modern electric living and, as the service symbol for the local utility company, he emphasizes the [utility] company's contribution to modern electric living so that its role is not lost in the appliance story."[25]

Companies that chose not to use Reddy as their mascot sometimes designed their own. In the 1950s, for example, Chicago's Commonwealth Edison invented "Little Bill," a likable tyke who emphasized with his name how unit costs for electricity had been declining (Figure 11). Another marketing tool, Bill also reflected the managers' pride in providing a useful product at ever-decreasing costs.[26]

The biggest concerted promotional push began in 1956 with the "Live Better Electrically" campaign. Conceived of by the General Electric Company (which manufactured generating equipment and home appliances), the "LBE" program constituted the first industrywide promotion since the 1930s.[27] In 1957, the LBE program incorporated the new "Medallion Home" program – an attempt to encourage sales of all-electric homes by placing a "visible but not conspicuous" medallion on the new house – and

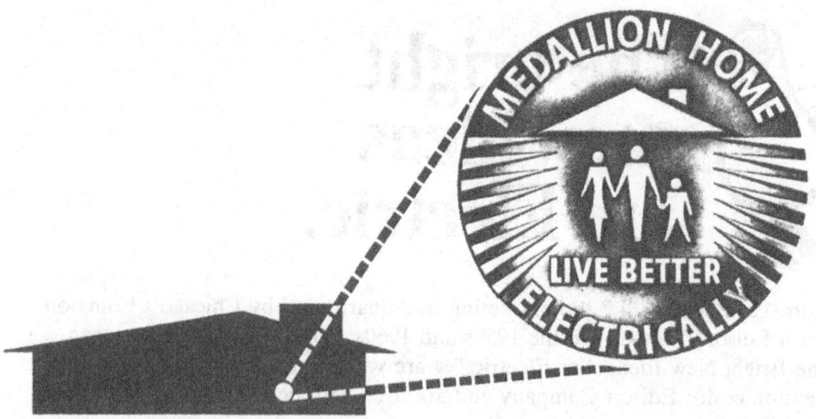

Figure 12. The Gold Medallion. Exhibited on all-electric homes, the medallion signified that the owners enjoyed a high standard of living. Reprinted with permission from McGraw-Hill, publisher of *Electrical World*, as appearing in Volume 148 *Electrical World*, October 28, 1957, p. 47.

by offering lower, promotional rates to their owners[28] (Figures 12 and 13). The program also sought participation from builders, realtors, and lending institutions, all of whom could encourage increased use of electricity. To help spread the word directly to the public, the program enlisted popular television personalities, such as Ronald Reagan, Betty Furness, and Fran Allison (Figure 14). At the same time, the federal government pressed its own efforts to sell the idea of safe nuclear-powered electricity.[29]

In short, utility managers vigorously pursued two goals concerning electricity usage. First, they felt an obligation to serve their customers and provide a means for improving their standard of living and economic welfare. They therefore needed to build new plants as the population increased and as enhanced industrial activity pushed up electricity usage. In doing this, managers simply responded to events occurring in a robust economy, and they saw their job as meeting that demand. But managers were not passive participants in a national phenomenon. They also aggressively pushed for increased usage of electricity, and they sought new ways to use and promote it. Utility journals exhorted "Sell or Die!" and "Sell – and Sell – and Sell"[30] while *Business Week* in 1956 reported on the extraordinary promotional efforts of the American Gas and Electric Company. The firm's incessant promotion of major residential loads – described as a "hard sell" – meant achieving saturations from 30 to 100% greater than the national average for new appliances.[31] The magazine underscored the importance of growth to AG&E by captioning the cover picture of the company president Philip Sporn with the words: "Doubling every 10 years is not enough."[32] Meanwhile, in a famous "Inventing Our Future" speech in

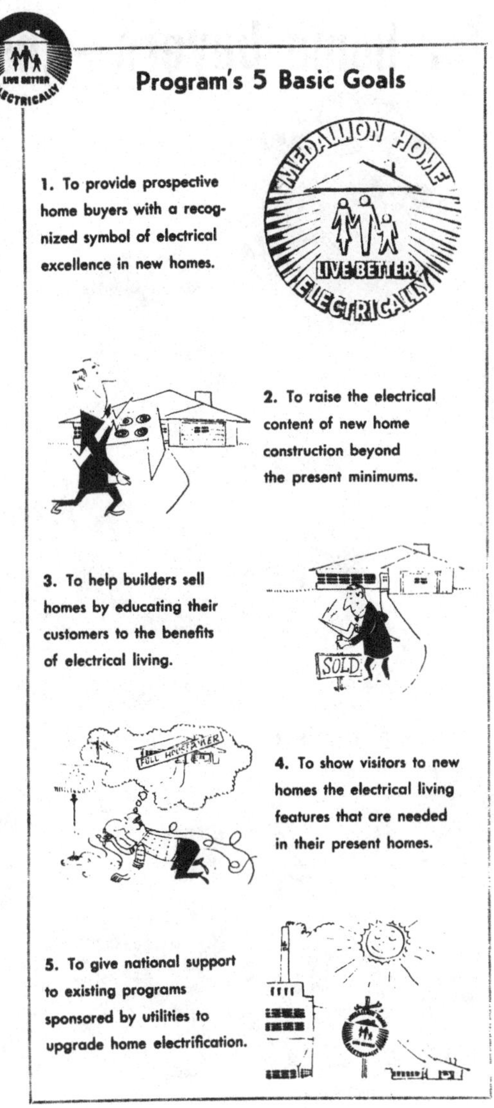

Program's 5 Basic Goals
MEDALLION HOME
LIVE BETTER ELECTRICALLY
1. To provide prospective home buyers with a recognized symbol of electrical excellence in new homes.
2. To raise the electrical content of new home construction beyond the present minimums.
3. To help builders sell homes by educating their customers to the benefits of electrical living.
SOLD
4. To show visitors to new homes the electrical living features that are needed in their present homes.
5. To give national support to existing programs sponsored by utilities to upgrade home electrification.

Figure 13. Basic goals of the Live Better Electrically, Gold Medallion Program. Reprinted with permission from McGraw-Hill, publisher of *Electrical World*, as appearing in Volume 148 *Electrical World*, October 28, 1957, p. 48.

1964, Sporn noted that "the most important elements that determine our loads are not those that happen, but those that we project – that we invent – in the broad sense of the term 'invention.' You have a control over such loads: you invent them, and then you can make plans for the best manner of meeting them."[33] Of course, as soon as utilities saw the results of

Newest guide for home buyers – the

YOU GET WONDERFUL FEATURES

ELECTRIC APPLIANCES. Mrs. Stanley Johnson, Arlington Heights, Ill.: "I just love our Medallion home – especially the kitchen. All these electric appliances that came with it – like this wall oven – sure make my job much easier. And my husband says they're easier to buy this way, because we pay for them on the mortgage."
178

LIGHT FOR LIVING. Mr. and Mrs. Charles R. McCarty, Greensboro, N. C.: "We never knew you could do so many beautiful things with lighting until we bought a Medallion home. Valance lighting, for example, makes our furniture and drapes look wonderful – and at the same time gives our son a well-lighted place to practice the piano."
BETTER HOMES & GARDENS, OCTOBER

Figure 14. Advertisement for the All-Electric Home, with Betty Furness, Ronald Reagan, and Fran Allison. Reproduced from *Better Homes and Gardens*, October 1958, with permission from the Edison Electric Institute.

Live Better Electrically MEDALLION

This new Medallion assures you a home has been inspected by the local electric utility... meets modern standards for wiring, appliances and lighting. Look for the Medallion. It means a wonderful new way of life for you and your family!

What Sterling is to silver... that's what this Medallion is to a new house! It's the new national symbol of the finest in electrical living. Let these three top TV stars, speaking here for the electrical industry, tell how you save trouble, time, and money by choosing a home that wears the Live Better Electrically Medallion.

BETTY: In a Medallion home, you start right off with a modern electric range, plus at least 3 additional major appliances, maybe more. They're installed, ready to go to work the day you move in! Appliances are easier to pay for this way.

RONNIE: The lighting in every Medallion home is specially planned, too. It provides better light for better sight, plus new beauty for your home. You also get full Housepower. This means enough power, wiring, circuits, switches, and outlets to handle all the appliances you want to use.

FRAN: You'll be glad all your life you bought a Medallion home. Read below what a few of the thousands of new Medallion home owners think of them. Then go see the Medallion homes in your neighborhood. Your electric utility will tell you where they are.

New Ideas for Better Living

The new Medallion is backed up by home builders, electric utilities, and electrical manufacturers (Frigidaire, General Electric, Hotpoint, Kelvinator, Thermador, Westinghouse, Whirlpool, and others). This year, utilities will award Medallions to [illegible] new homes — in every style and price range across the country. You'll see lots of new ideas in the Medallion homes on display now!

LIKE THESE IN MEDALLION HOMES!

FULL HOUSEPOWER. Mrs. Nick Piscopiello, Meriden, Conn.: "One of the things I like most in my Medallion home is all the handy outlets. I can plug in my portable cooking appliances wherever I want and use them — even with the washer going — without ever blowing a fuse. And I can cook a meal anywhere in the house — and outdoors, too."

ELECTRIC HEATING. Many Medallion homes feature electric [illegible] too. These are awarded a special Gold Medallion. The [illegible] pump, shown here in the home of Mr. and Mrs. [illegible] of Beverly Hills, California, provides year-round comfort [illegible] unit which automatically heats or cools as the weather [illegible]

:79

...TTER HOMES & GARDENS, OCTOBER, 1958

their sales efforts, they then needed to build new plants to meet demand. And so the spiral continued into the 1950s and 1960s.[34]

Utilities' pressure to scale up technology

As befitted the basic tenets of the grow-and-build strategy, load growth meant that utilities needed to install larger and more efficient power equipment. But unlike the situation before World War II, utilities desired – and manufacturers produced – technology that emphasized scale improvements more than thermal-efficiency gains. While the best units improved in efficiency from about 32 to 40% in the postwar period,[35] single-shaft turbine unit sizes jumped from about 160 to over 1,000 MW. When New York's Consolidated Edison Company installed its huge 1,000-MW unit in 1965, for example, the machine exceeded the capacity of the previous biggest by 323 MW, installed just two years before. The size of this increment is significant when considering that it took the industry seventy-four years (to 1956) just to install its first unit of 300 MW.[36]

The continued push for larger units had several causes. First, growth rates in electricity usage appeared to be soaring, and utilities saw large units as one way to keep up with demand. As the country enjoyed postwar economic prosperity, consumers spent money on automobiles, televisions, and other "luxuries" as never before. Many of these, such as air conditioners and electrical space-heating systems, helped push electricity consumption up almost beyond belief. In 1955, managers marveled at a 17% leap in electricity sales from the year before. This jump followed other big years, making for an average annual growth rate of 10.8% for the immediate post-World War II decade. While annual sales growth moderated in the late 1950s and 1960s to a more "traditional" 7 to 8% range, the years still were punctuated by spurts of 10.1% in 1959, and 8+% figures in 1966, 1968, and 1969.[37]

But even these aggregated national figures hide the exceptional growth rates for specific regions, such as in the booming south central states. Serving the growing "Virginia Urban Corridor" between Washington, D.C. and Richmond, Virginia, the Virginia Electric and Power Company watched its load growth soar as the region's population growth rate zoomed 2.5 times faster than the national average. Between 1958 and 1959, the company's usage rate increased 12.3%, while its peak load (the maximum use for the year) grew at a 12.4% rate – meaning a doubling period of only six years if the rate continued. Almost a decade later – in 1968 – the company witnessed even more spectacular growth rates of 12.6% for sales and 21.6% for peak load.[38]

Other utilities also needed to meet increasing demands while assuring that they had enough power – and reliable power – especially after the November 1965 blackout in the northeast.[39] The event spurred firms to reevaluate plans for installing new capacity to ensure reliability, while at

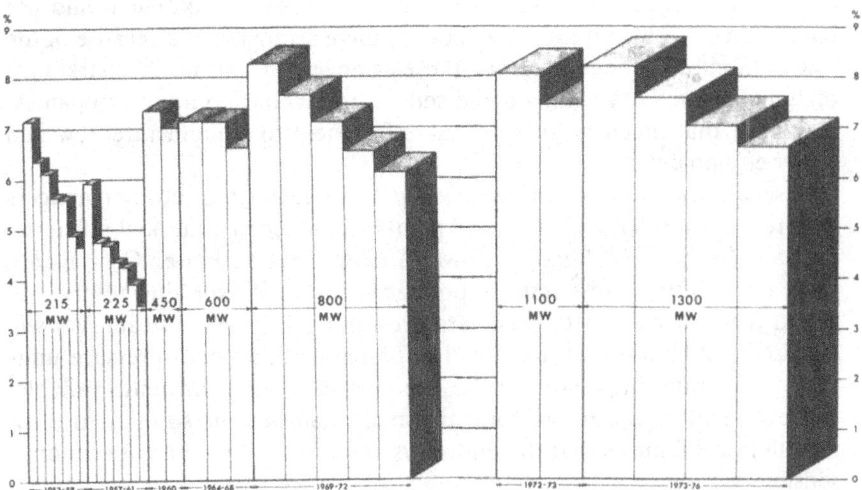

Figure 15. Capacity additions to the AEP system. Illustrating the "7% to 10%" rule of thumb, the graph shows how the AEP constructed new units that never exceeded about 8.5% of its system size. In the period between 1973 and 1976, for example, the company built four units in the 1,300-MW range, the last of which represented an addition of about 7% to the company's system capacity. Reprinted with permission by the American Electric Power Company.

the same time increasing the need to interconnect safely with other utilities. The blackout led to the creation of regional power pool arrangements through the new North American Electric Reliability Council. Establishing nine regional pools for ensuring reliability of the entire network, the industry-formed council coordinated operations by integrating small utilities into larger systems. The groups of companies enabled purchase of bigger units because of the the "7 to 10%" rule of thumb for capacity additions. Dating to the early part of the century, the rule limited a company from installing a new unit that accounted for no more than 7 to 10% of the entire system's capacity.[40] Adherence to the rule guaranteed that companies could provide reliable service even if the new plant went out of service unexpectedly. A plant larger than 10% of system size might strain the reserve margin and cause blackouts. As noted in 1969 by John Tillinghast, executive vice president of engineering and construction for the American Electric Power Company, the "addition of supersized generating units sometimes is like putting too many eggs in one basket – if the unit is in trouble, the entire system is in trouble"[41] (Figure 15).

Because these pooling arrangements effectively increased the size of power systems, companies could buy larger units than their individual sizes would dictate. Thus, small utilities such as the Public Service Company of New Hampshire, working with other firms in its region, could consider building large nuclear power plants in the 1960s – plants that would not have been conceivable earlier without pooling arrangements because of the rule of thumb. To manufacturers, these arrangements meant the market for big units grew. They simply could sell their machines to more companies, making it that much more reasonable for them to manufacture new and larger equipment.[42]

A second pressure for more powerful units came from utility managers who recognized that existing fossil plants had reached a thermal-efficiency plateau. In the 1964 *National Power Survey*, Federal Power Commission staff and utility representatives pointed out that diminishing returns resulted from increasing the temperatures and pressures of steam. As will be discussed later in greater detail, high-temperature and -pressure units then in operation appeared to be the most thermally efficient available. But even with them, the metal components had become so thick to resist metallurgical failures that the entire system pushed the limits of economic sensibility.[43]

To continue improving the productivity of the utility industry *without* augmenting thermal efficiency therefore became a management goal. Traditionally, efficiency enhancements went hand in hand with gaining economies of scale. But now one course of improvement had been eliminated. Beginning in the early 1960s, then, companies that added new plants took advantage of prototype units that offered greater capacities than previous ones. Managers realized that the units' thermal efficiency would not grow and that the only way to continue reducing the cost of producing electricity was twofold: by replacing much older plants (which had lower efficiencies than the current versions), and by exploiting economies of scale in the new plants[44] (Figure 16). By 1968, the trend seemed clear, with *Electrical World* reporting "that gigantic generating units are the key to low bus-bar energy costs. . . . [S]tation designers seem to have decided that more can be gained by exploiting the economy of scale than by further efforts to improve the [steam] cycle efficiency."[45]

While managers may have lamented the loss of thermal efficiency improvements, they were heartened by a countervailing fact: at the same time that efficiency reached a plateau, the cost of fuel declined. On average for the nation, fuel costs declined 10% in current terms and 24% in constant terms from 1957 to 1966.[46] The cause of the decline consisted of a generally stable – if not declining – coal industry that suffered as manufacturing and transportation industries shifted to oil, which also declined in price as foreign supplies flooded the U.S. market.[47] Of course, the electric utility industry still used coal, doubling its use between 1945 and 1955, but large surpluses remained.[48] As a result of these fuel-market shifts and the prospect of still cheaper-to-operate nuclear-powered plants, utility managers did not fear the loss of thermal-efficiency improvements.

Inflation and the desire to continue achieving lower costs despite it also constituted prime motivating factors toward more powerful units. While

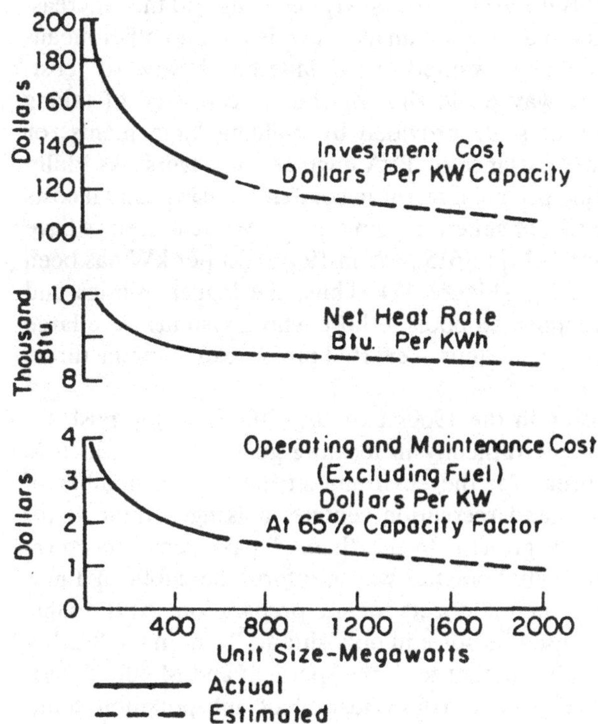

Figure 16. Performance and cost characteristics of coal-fired steam electric plants. The figures demonstrate how cost per kilowatt of new capacity decreased as units increased in output. Thermal efficiency, as noted in the second graph, however, did not change appreciably after reaching the 400-MW size. Reprinted from Federal Power Commission, *National Power Survey* (Washington, DC: U.S. Government Printing Office, 1964), p. 70.

today's electricity consumers may be skeptical of the concerns of utility companies, the firms' managers took justifiable pride in providing cheap and abundant electricity supplies. But with inflation accelerating during the Korean War, the 1956–8 period, and the late 1960s, costs of construction became harder to contain. As described by the Handy-Whitman index, a measure of the costs of constructing generating stations, building expenses increased 53% from 1950 to 1959. Between 1959 and 1963, the index actually declined 1%, only to jump 24% between 1963 and 1969 – followed by a 48% increase in the first five years of the 1970s.[49] Meanwhile, after a period of eleven years when the nominal cost of fossil fuel declined by more than 10%, fuel costs turned around in the late 1960s due to shortages of coal and mine-worker strikes in parts of the country. On average, these costs rose 18.7% in nominal terms between 1966 through 1970.[50] To an organization like the TVA, cost cutting had become a "crusade" – in the words of one manager.[51] But the effect of slowly declining and then increasing prices for fuel dictated that utility managers needed to find other means to improve productivity if they wanted to continue the downward "cost-price sales spiral."[52] One way to do this, of course, consisted of taking advantage of economies of scale provided by building large plants, or "kilowatt factories" in the parlance of TVA engineer-managers. As Philip Sporn put it in 1964, "[t]he principal factor in combating rising capital costs has been the increase in size of generating units. . . . While unit sizes have risen about six-fold [since 1929] to 615 MW in 1964, cost per kW has been reduced to about half . . ."[53] (Figure 17). Thus, the bigger utilities and members of large, interconnected pools – those whose systems were large enough to take on high-capacity units – exerted pressure on manufacturers for ever larger units.

Another problem arising in the 1960s that contributed to the push for larger units consisted of the difficulty in locating good sites on which to build plants. Public concern for the environment limited the number of locations near population and recreation centers, making it difficult for utility managers to plan for growth. In the "happy" past, reported *Power Engineering* magazine in 1970, "practice was to inform the public of a new power plant by moving in the bulldozers for site preparation. At the same time, the public seldom raised its voice in opposition."[54] But in the "turbulent 1960s," when "the environment took the spotlight and Middle America joined with activist youths in a groundswell of anti-pollution sentiment,"[55] managers concluded that once a good site had been located, they should build the largest possible plant. The TVA and other large systems encountered the siting problem earlier than most others. But the huge size of the systems also meant that they would suffer the least reliability penalty when installing a big power unit (according to the 7-to-10% rule). As a result, the siting problem provided another good reason for big utilities to install the most powerful unit possible.[56]

A final pressure for larger units came from the burgeoning nuclear power industry. The new source of electricity appeared attractive to many utility managers, especially to those in resource-poor New England, who looked to nuclear power as a way to avoid dependency on oil and coal. But to prove financially viable, nuclear plants had to be enlarged dramatically so that scale economies would provide cost advantages over conventional power plants – a major goal of the nuclear manufacturers. In a strange way, too, the lower costs of fossil fuel in the early 1960s stimulated the tendency to build larger nuclear units, so that they could be still more competitive with conventional units.[57] Within the skewed market of government subsidies and loss-leader pricing from manufacturers, then, nuclear units "grew" very rapidly in the 1960s to take advantage of scale economies arising from their construction. In less than a decade after first being introduced, nuclear units reached about the same size as their fossil counterparts.[58]

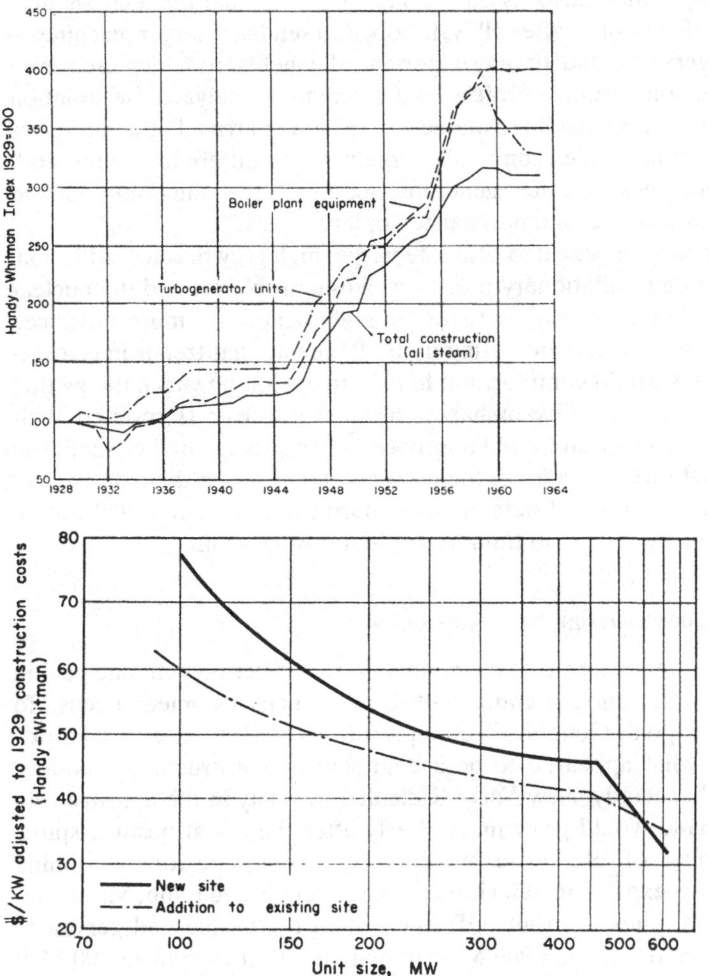

Figure 17. Graphs illustrating AEP's experience with growing construction expenses but lower unit capital costs. The scale economies arising from more powerful units were largely responsible for this trend. Reprinted with permission from Philip Sporn, *Vistas in Electric Power*, Vol. 1 (Oxford: Pergamon Press, 1968), p. 194.

Good evidence existed to support the value of scale economies with fossil-fuel technology. "Common knowledge" and experience from years of practice in the early twentieth century provided one source of information, as was noted in the previous chapter. In addition, the 1950s witnessed the beginning of a flurry of studies that supported the common view on scale economies. One engineering study in the 1950s, performed by General Electric engineers, demonstrated dramatic decreases in investment, maintenance, and operating costs as output increased.[59] While the analysis may have been self-serving – after all, GE hoped to sell these larger machines – utility managers who had already experienced benefits from increased scale seconded the conclusion.[60] Professional economists analyzed the situation too, using their specialized techniques. They never arrived at a consensus on how to optimize scale economies, largely due to differences in methodological approaches, but they generally sided with the conclusion derived from experience that economies existed in large units.[61]

All these concerns – zeal to promote usage, high growth rates, a thermal efficiency plateau, inflationary pressures, siting problems, and the nuclear bandwagon – led to pressures to order higher-capacity, more advanced fossil and nuclear steam-generating units. Believing that trends in electricity growth rates would continue and hoping to exploit new technology that would reduce costs, utility managers after World War II restated their allegiance to the grow-and-build approach.[62] The strategy had worked well during the first three decades of the century, and now that the industry had returned to a more normal state of affairs during the 1950s and 1960s, there appeared to be no reason to think it would not work again.

Vendors' response: Design by extrapolation

The manufacturers entered the immediate postwar period with one goal in mind – to produce enough equipment to meet their customers' needs. To accomplish the goal, General Electric president Charles "Electric Charlie" Wilson took what appeared to be a bold step by constructing a modern factory in Schenectady, New York. Ridiculed by many in the industry who felt that demand would grow more slowly after the initial postwar spurt, "Wilson's Folly" of 1949 soon proved to be inadequate for the coming orders, and new expansion of facilities in Schenectady and Lynn, Massachusetts, followed quickly. In all, GE augmented its turbine and generator production facilities from 3,000 MW annual output in 1946 to 12,500 MW in 1963.[63] Though Westinghouse and Allis-Chalmers also built new facilities to meet growing demand, GE maintained its prewar market share of about 60%. The other two firms divided the rest of the market in the 1950s and early 1960s with about 30 and 10% shares, respectively.[64]

New factories took time to build, and immediately after the war, the manufacturers developed a short-term strategy to help meet capacity shortages without them. Cooperating with a trade association that established industry standards, the vendors agreed to give up their quest to make major innovations with each pioneering machine. Instead, they decided to produce quickly manufactured units of fixed capacities.[65] The approach proved successful for alleviating the power shortage. But even after utilities reestablished safe reserve margins, the manufacturing approach still appealed to Westinghouse, which produced 54% of its equipment in standard designs in 1950. Unfortunately for the company, utilities in the early 1950s began demanding larger, custom-designed units that operated optimally within their specific systems, and GE offered them. Westinghouse responded to GE's divergence from the accepted strategy by instituting a crash program to develop competitive machines. The transition to a new production approach strained the company's resources, which were devoted to providing new design equipment for recently hired (and overworked) engineers.[66]

Though the utilities' postwar grow-and-build strategy appeared similar to the traditional and successful approach used earlier in the century, the extra emphasis on large scale in the 1950s and 1960s put new pressures on manufacturers to develop a novel design philosophy. One element of it consisted of foregoing some of the conservatism in the design-by-experience approach by planning new and bigger machines before practical knowledge had accumulated from previous units. Called "design by extrapolation" by some people in the industry, the new approach differed dramatically from its predecessor. Instead of waiting for experience – and learning – to accrue while observing how earlier machines operated in the field, manufacturers made abrupt jumps in design to the next stage. "The prewar practice of advancing steam temperatures and pressures only after the industry had acquired an adequate fund of knowledge and experience gave way," reported the authors of the 1964 *National Power Survey*. "New and larger units employing higher temperatures and pressures were ordered even before the lessons could be learned from the previous advances."[67] Using the method, for example, manufacturers would design a 300-MW turbine generator just like a 200-MW machine, except with scaled-up components and with little time intervening between deployment of the two units.[68] One Westinghouse engineer noted that the company extrapolated "on the basis of calculations, model tests, and the belief that if a small step is ok, the next one would be ok. . . . But before the experience on that design came to pass, we were into building lots of them, and the next step was already in place."[69]

Manufacturers adopted the approach because it yielded rapid advances in the capacities of components – exactly what utilities demanded. But vendors benefited too by developing increasingly powerful turbines and generators – equipment for which they could earn good profits. General Electric, for example, developed a policy of pricing equipment based on its value to utilities rather than on the basis of its cost plus a specified percent profit. Because of economies of scale inherent in more powerful units, utilities valued large units as a way to reduce their capital cost per kilowatt and as a way to deal economically with increased demand. At the same time, however, productivity improvements in the manufacturing process brought down the cost of manufacturing the equipment. Thus, General Electric could offer its new, higher-capacity equipment at a unit price that remained lower than previously sold machinery, but still considerably higher than its direct production costs. In such a way, both the utility customers and manufacturer shared the fruits of technological improvements. Since unit costs continued to decline, vendors experienced little resistance from customers, and according to one academic observer of the industry, "with rising value, prices could soar."[70] Westinghouse appeared to use the same value-based pricing scheme. One manager indicated that the firm always preferred to sell its large-capacity, advanced equipment because the profit margins were higher on the big units.[71]

Meanwhile, the manufacturers and utilities felt confident that forsaking the design-by-experience approach would still lead to successful technology. After all, designers since World War II had new materials at hand – those that were stronger and more resistant to aggressive steam conditions – and in the 1960s, they had begun to use computer-based analytical design tools. Moreover, they realized that the design-by-experience approach often produced equipment that performed better than necessary. In other words, manufacturers often overengineered machines by building them very conservatively and with large margins for error. Because the next version would be based on the previous success, it too would contain lots of margin. Consequently, utilities often obtained an unexpected bonus in performance of their new equipment. As an example, Louisiana's Gulf States Utilities Company benefited in 1960 by installing a unit rated at 130.5 MW. Because of the extra margin, the company operated it reliably at 162 MW, receiving 24% more power than bargained for![72] Another "happy" customer that profited from the design practice was the TVA, which bought several units rated at 150 MW and then rerated them at 175 MW.[73] The existence of this extra margin helped convince managers that extrapolations into unknown territories with larger-capacity units would not be especially risky. Through the 1950s and early 1960s, it appeared that the approach proved successful. "[T]hings were working," claimed one Westinghouse engineer. "We thought we understood what we were doing."[74]

Finally, the design-by-extrapolation approach appealed to some manufacturers as a way to compete and demonstrate technical superiority. Consider, for example, the case of Allis-Chalmers, which in the early 1960s attempted to "leapfrog" previous designs and build a 1,000-MW power unit for New York City's Consolidated Edison Company. At the time of the order's announcement in 1961, the largest unit in use produced about 600 MW, whereas the most powerful made by Allis-Chalmers was smaller than 400 MW.[75] Nevertheless, manufacturing and utility managers felt that extrapolation to huge capacity could be made, with 1,000 MW being a "magic number."[76] Moreover, the large unit would provide tremendous economies: when installed in 1965, the machine cost 30% less per kilowatt than a 400 MW unit.[77] Along with a big sale, the company would also win the respect of the utility industry, which would perhaps now view the former third-place company as a technological leader.

Manufacturers' use of the new approach: Rapid scale increases

Manufacturers exploited the design-by-extrapolation approach to the fullest in order to effect rapid scale increases in all power-plant components. To *boiler* manufacturers, this meant that they needed to produce ever greater steam volumes to power the growing turbines. The efforts generally proved successful. In 1946, for example, the most advanced conventional boiler produced 1 million pounds of steam per hour; by 1962, the best machine yielded over 6 million pounds per hour[78] (Figure 18). During the same period, the average capacity of boilers grew from about 500,000 pounds per hour to about 1.5 million pounds.[79] More impressive were the so-called "universal-pressure" boilers, which could produce steam at above or below the point where water turns to steam without boiling (the critical value). Operated in 1957, the first such boiler delivered 675,000 pounds of steam per hour for a 120-MW turbine generator.[80]

Though a necessary component in increasingly large power plants, "growing" boilers did not contribute as much to reduced capital costs per kilowatt as did turbine-generator sets. According to one econometric study of 126 power plants built in the United States between 1948 and 1963, 63% of the reduction in unit capital costs derived from scale economies, with the most important scale variable consisting of the turbine-generator capacity. Likewise, the power output of the turbine generator proved to be the major determinant in the decreasing cost of labor.[81]

Considering each turbine-generator component separately, we see that more powerful *turbines* could usually be produced by two methods, often used together in one large machine. First, manufacturers designed larger turbine blades for use in the low-pressure stages of turbines. These "buckets" (in GE parlance, "blades" to Westinghouse engineers) withdrew extra energy before the steam exited the turbine. Secondly, extra low-pressure turbine casings (attached to the same shaft and consisting of blades and associated hardware) could be added when appropriate to take advantage of energy left in steam before it exhausted into the condenser. Additional low-pressure casings provided a larger exhaust area and led to greater efficiency because better vacuums resulted.[82] As an example, a large 1,000-MW nuclear turbine could consist of four separate casings housing low-pressure blades as long as 43 inches.[83] A fossil unit of similar capacity might employ 33.5-inch long blades in five casings.[84]

Generator cooling offered the greatest hindrance to larger-scale equipment before World War II. But after the conflict, sustained efforts to invent more efficient techniques for removing heat produced in the machine's coils of wire led to the use of liquids previously employed in transformers, with which power engineers had much experience. Still better, though, was water, which promised a heat-removal capacity fifty times greater than air. Improved methods of channeling the heat-removing liquid through the machine also aided cooling. Of course, new technical challenges arose as

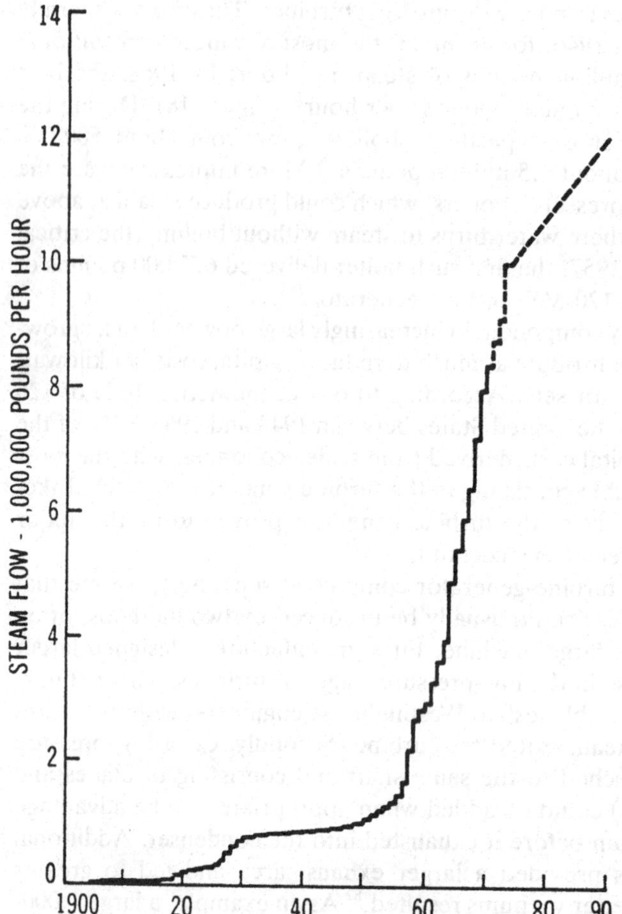

Figure 18. Maximum capacity of boilers installed in each year. The graph demonstrates how boilers produced increasing amounts of steam to make possible the growing capacities of steam turbine-generator units. Reprinted from Federal Power Commission, *1970 National Power Survey*, Vol. 1 (Washington, DC: U.S. Government Printing Office, 1971), p. I-5–11.

generators grew. Problems with insulation, vibration in conductors, short-circuit forces, and quality assurance on large field forgings required resolution as new cooling techniques permitted higher output from generators.[85]

Cooling advances eliminated the need to use turbines with two shafts and

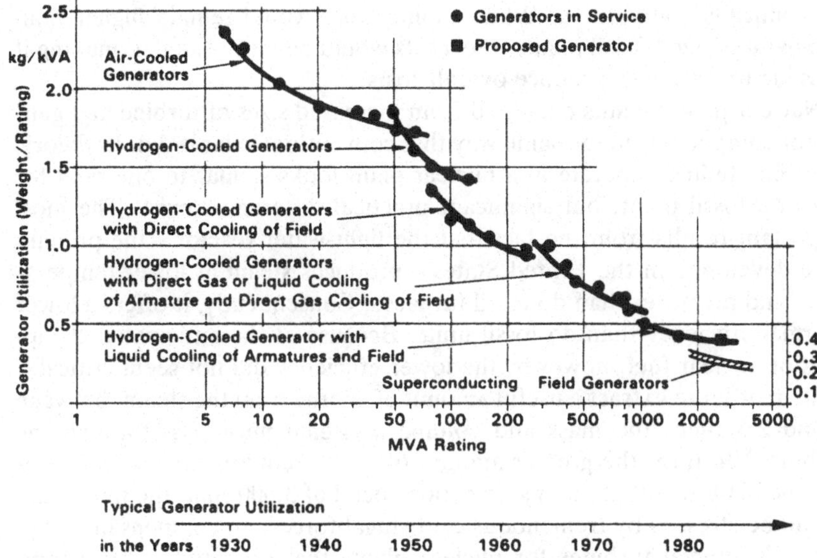

Figure 19. Reduction in weight/rating of generators. Engineers used different cooling techniques to reduce the amount of metal needed to produce electricity in generators. This made it possible to increase the electrical output of generators (and reduce their cost) so that a single large turbine could turn a single generator. Reprinted with permission from Vol. 39, *Proceedings of the American Power Conference* (1977), p. 257.

two generators (cross-compound units) for obtaining greater power capacity. Now, using only one single-shaft turbine and an efficiently cooled generator, manufacturers could make more compact, yet more powerful units. General Electric proudly proclaimed in 1953 that the "remarkable achievement [of liquid cooling] will permit much larger ratings of generators – in the same frame size – and opens the way to increased size of the 3,600-rpm tandem-compound turbine-generator set."[86] And the prediction came true with startling productivity improvements, as measured by the growth in "generator density" – the ratio of a generator's weight per kilovolt-ampere (kVA) of power produced. Between 1940, when utilities commonly bought air-cooled machines and 1975, when they used water-cooled machines, the density declined from 5.5 to 1.0 pound per kVA[87] (Figure 19). As a result, single units that combined turbines and generators rose to over 1,000 MW in the early 1970s with significant economies from producing all power on one shaft.

Scale increases of components also accompanied the growth of the nuclear power industry. After the first demonstration reactors had been built, utilities started a period of frenzied ordering of new plants in the early 1960s. While experience had not proven that nuclear power would be cheaper in terms of capital costs than fossil plants, both manufacturers and utilities believed that economies of scale in *large* units would make them economically competitive. While capital costs would remain higher than comparably sized fossil plants, fuel costs would be much smaller, making it possible for utilities to reduce overall costs.

Nuclear power plants benefited from increased sizes of turbine and generator components in the same way that conventional plants did. In theory, a turbine built to operate in a nuclear plant looks similar to one built for use in a fossil plant, but significant practical differences exist. The most important results from the fact that the light-water reactor – the primary type developed in the United States – produces steam at lower temperatures and pressures than do fossil furnaces. Consequently, it offers a lower thermal efficiency than do fossil units. Because of the anticipated cheapness of nuclear fuel, however, the lower efficiency did not seem critical.[88] Still, in order to extract a useful amount of energy from the steam, between 2 and 2.5 times the mass and volume of steam must pass through the turbine.[89] In turn, the greater amount of steam requires the use of longer turbine blades. But at the high rotation speed of 3,600 rpm, the tall blades would be stressed by tremendous centrifugal forces. To compensate, engineers developed turbines for nuclear plants that operated at 1800 rpm, returning to modified designs that had been used before World War II. However, engineers also took advantage of the new metallurgy, computer-aided designs, and other advances since the war to produce modern versions of "slow" turbines.

Boilers, of course, differed greatly in nuclear stations. But the special physical characteristics of the cylindrical nuclear core made leaps in size very attractive. As the radius of the core increased linearly, the volume of steam produced increased as the square of the radius (given the same depth of cylinder). Meanwhile, because of the low relative cost of uranium, fuel costs are not greatly affected by unit size.[90]

"Balance-of-plant" considerations also suggested building larger nuclear power plants. Unlike fossil plants, nuclear facilities require safety measures to protect against leakage of radioactive water and gases. Duplicate emergency cooling systems constituted one method of protection; another consisted of enclosing the entire boiler and heat-transfer mechanisms in a large containment building – basically a concrete dome. Partly mandated by the Atomic Energy Commission as safety and reliability assurances, these capital additions meant extra costs, but they did not grow proportionally with a unit's size. Therefore, managers believed that the fixed costs of a nuclear plant per unit of power would decrease more rapidly than those of a fossil plant.[91] Summarizing this view, the 1964 Federal Power Commission survey noted that "the capital cost disadvantage of nuclear plants in comparison with conventional plants will be less significant for larger plants."[92]

Largely because of these perceived economies of scale (and incentives from the federal government and manufacturers), nuclear power units grew rapidly. The first commercial unit, built in 1957 at Shippingport, Pennsylvania, had a rating of 60 MW. By 1960, the next pressurized-water

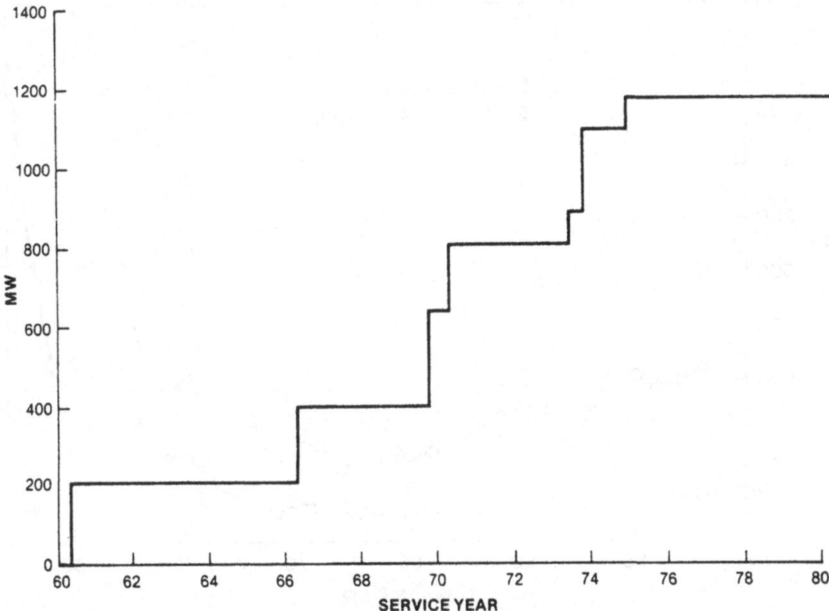

Figure 20. Growth in megawatt rating of GE nuclear units vs. service year. The capacity of nuclear units grew very rapidly during the "bandwagon" years of the 1960s. Reprinted with permission from the General Electric Company as appearing in R. C. Spencer, "Evolution in the Design of Large Steam Turbine-Generators," General Electric Large Steam Turbine Seminar, Paper 83T1, 1983, p. 3.

reactor of 175 MW had been constructed. As of July 1964, similar units having capacities of 375 MW had begun to be constructed.[93] And in a remarkable demonstration of faith in the design-by-extrapolation method, utilities in 1966 ordered nuclear units as large as 1,090 MW, even though manufacturers had no practical knowledge at the time with units bigger than 200 MW[94] (Figure 20).

Overall, the novel manufacturing approach for effecting rapid scale increases in electric-power technology (conventional and nuclear) did what it was supposed to do: produce cheaper and more powerful units. In 1950, when the average capacity of installed units stood at 26 MW, new construction costs amounted to $173 per kilowatt. By 1965, when the average output reached 103 MW, the cost of a new conventional plant declined to $101 per kilowatt in current terms, despite substantial cumulative inflation in the intervening years[95] (Figure 21). Lower cost was not the only benefit, however. Advances in scale enabled the industry to sustain its growth with fewer plants. Between 1956 and 1970, for example, utilities operated 58 fewer plants to produce 179% more electricity.[96] These triumphs signaled progress, and few people within the industry could dispute it!

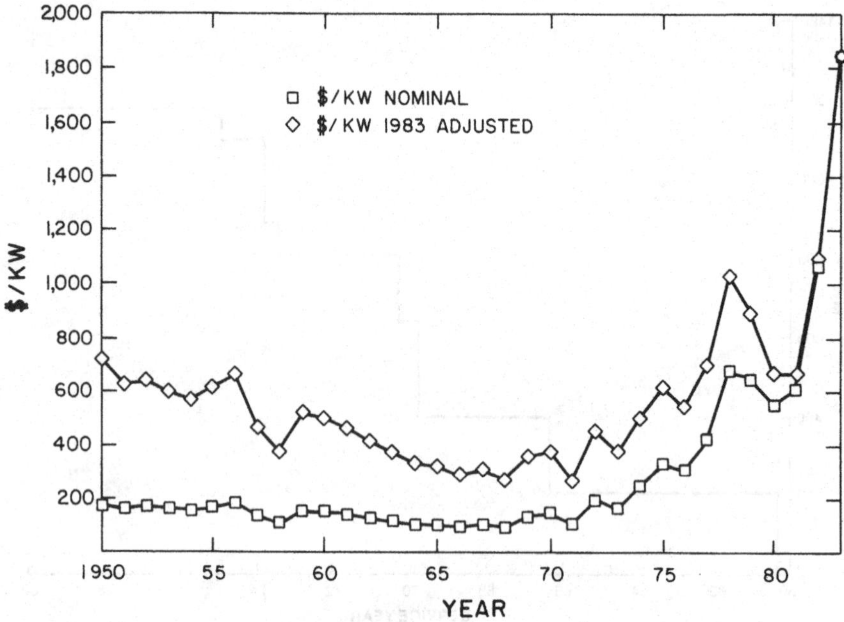

Figure 21. Incremental cost of generating plants. The cost of building a new plant on a per-unit basis decreased until the late 1960s, despite increases in the cost of almost all materials and labor. Exploitation of larger power units that demonstrated scale economies was responsible for the drop. After 1967, building expenses increased dramatically, contributing to the reversal of the industry's trend toward declining overall costs. Data are from Leonard S. Hyman, *America's Electric Utilities: Past, Present, and Future* (Arlington, VA: Public Utilities Reports, 1983).

The design-by-extrapolation approach, therefore, appeared to be successful into the mid-1960s. Larger-scale technology that emerged from manufacturers' factories in which the new approach had been employed provided productivity enhancements that utility managers demanded. To be sure, the utilities gave up some of their previous caution concerning their use of only well-tested technology or modest enlargements of smaller-scale equipment (as produced using the design-by-experience approach). But events during the postwar era differed from what managers had seen earlier in the century. Besides escalating costs and demand for electricity, utility managers needed to deal with the effective plateau of thermal efficiency and other pressures that promoted the rapid scale-up of equipment. Hence, utilities accepted what appeared to be only minor risks inherent in a new manufacturing strategy pursued by their traditionally reliable manufacturers – a strategy that seemed to offer universal benefits.

5
Utilities' role in technological progress

To explain progress in electric power technology simply as a result of research and development performed by manufacturers would be one-sided and misleading. Managers in several utility companies also played major roles in stimulating advances. Not just the complacent buyers of new technology, managers in these utilities often pressured manufacturers to develop advanced equipment, and they offered their companies as "guinea pigs" for untested machinery. Their leadership role enabled manufacturers to design new equipment and have it evaluated in the field, thus allowing them to gain experience. The technology could then be exploited by the vast number of "follower" utility companies.

To understand the role of utilities, one needs to understand how utility managers selected and made decisions about technologies. This in turn requires an examination of the values and community traditions shared by the managers, who more often than not had engineering backgrounds. Because of their technical training, high-level executives cherished values that differed from managers who understood only "dollars and cents" concerns. They hoped for and worked toward financial success for their companies, but they did not always see this as their primary goal. Big, "neat," and exciting technologies often caught their imagination, distracting them from purely economic considerations. In short, a dichotomous set of goals characterized their decision-making process: the quest for technical superiority as well as financial success. Their devotion to this set of goals helps explain not only why utility managers aided manufacturers in developing new technology. It also suggests why they managed their companies differently from people with backgrounds in accounting or finance.

Community traditions and technical values

On the surface, the goals of engineers and managers appear divergent. After all, the engineer is conceived of (in the public mind at least) as "a skillful manipulator of physical substances and the organizer of vast technical enterprises."[1] He strives for efficient use of natural resources in order to serve the material needs of humanity. In contrast, managers conform to the goals of the institution in which they work. They attempt to organize, motivate, and supervise people so that the institution reaches established goals (such as profitability for most corporations, and aid to various constituencies for nonprofit or government organizations). As one observer put it, "engineers deal with things, managers deal with people."[2]

Despite this seeming conflict, engineers have successfully blended their own objectives with those of their institutions. In most industries, engineers embraced the goals of corporate capitalism to such an extent that considerations of cost and profit became as important as knowledge of the physical properties of matter.[3] Reflecting this view in 1903, Charles F. Scott, president of the American Institute of Electrical Engineers, expanded the notion of engineering to encompass the "*art of organizing and directing men*, and of controlling the forces and materials of nature for the benefit of the human race."[4]

In the electric utility industry, where large-scale technology used for generating, transmitting, and distributing electricity required technically trained people, it is no surprise that many engineers joined companies. But beyond serving as operators of machinery, engineers earned their way into management ranks. Samuel Insull, for example, recognized the importance of employing trained engineers for managing large and growing utility companies. Though Insull himself did not have a formal technical background, he implored young engineers in universities to learn more about the commercial realm of utility work. Speaking before colleagues at the 1910 National Electric Light Association convention, he argued that if "we are to make relatively as great progress in the development of the central-station business as has been made in the last quarter of a century, it will be necessary for the technical institutions of the country to give greater prominence to the commercial side of central-station business and . . . to teach them more of the true conditions governing commercial development."[5] The usefulness of engineers in management of technology-based industries also won support within the American Institute of Electrical Engineers. In the years before World War I, Institute members even coined a new expression – "commercial engineering" – to describe the activity in which engineers managed a business and not just a system of machines.[6]

Efforts such as those made by Insull and professional engineering societies bore fruit in bringing an abundance of engineers into the management of utilities. Throughout the twentieth century, utility managers acquired their practical or educational training as engineers, working their way up the corporate ladder from entry level technical positions or having come from universities after graduating. In the period after World War II, managers continued to obtain similar training. When asked by the author about their formal backgrounds, a great majority of utility executives indicated that they had been taught to be engineers, as were their predecessors earlier in the century. Independent sources support this finding: a 1964 survey indicated that 67% of utility management personnel were engineers, as were two-thirds of all board chairmen and company presidents.[7] A more recent study of fifty-three managers in two utilities noted that 64% had college engineering degrees.[8]

When trained as engineers, these people developed a sense of community traditions that they were unlikely to discard after becoming managers. As noted by Edward Constant in his study of aircraft engineers, these traditions included a knowledge dimension, which came from an understanding of basic engineering principles and practice. This background provided a standard approach toward solving problems – the "engineering method" – that attempted to isolate a system and control its variables. The tradition of practice also included a sociological dimension and behavioral norms.[9] In other words, utility engineer-managers, like other engineers, shared the basic community values and goals of the engineering profession in the United States. Among these values was the enjoyment of using advanced forms of technologies that promised to push back previous barriers to improved performance. When successful, the utility engineer-manager could boast at national meetings, thereby gaining the prestige of his colleagues, and he enjoyed reading about his firm's technological prowess in industry journals. No doubt, he also made a point of informing stockholders and financial analysts about the company's technological achievements, which carried implications of effective overall management.

Actions that date to early in the century provide evidence for the subordination of simple management efficiency (i.e., profitability) to engineering community values that embraced the goal of technological leadership. To engineers outside the utility industry, many of the "radical" steps taken by "aggressive" managers may have appeared conservative by comparison. Instead of adopting totally novel generating technologies, they generally depended on manufacturers who used the design-by-experience approach and who produced variants of existing technologies. Nevertheless, within the industry, a few managers departed from the norm and willingly took modest financial risks when using pioneering machinery in return for industry prestige. For example, the Edison Electric Illuminating Company of Boston boasted of its technical achievements in 1925 when it installed the first central station that performed at 25.1% thermal efficiency (heat rate of 13,600 Btu/kWh) through the use of steam at 1,200 pounds per square inch (psi). At the time, the average thermal efficiency of power plants stood at 15.5% (a heat rate of about 22,000 Btu/kWh).[10] In an introduction to an *Electrical World* article announcing the plant's operation, the company's president, C. L. Edgar, pointed out that general utility practice had been to use steam pressures of less than 400 psi, while a few firms experimented with pressures up to 600 psi. "We decided to take a bold step and go in one jump to 1,200 lb," he proclaimed. The successful use of high-pressure steam "is a credit to the company, to its designers and to its constructors."[11]

The desire to become the technological leader in the utility industry spurred others as well. Among the leaders in the 1920s, the Detroit Edison Company won accolades when it employed steam temperatures up to 1,000 degrees at a time when standard practice suggested a limit of 750 degrees. *Electrical World* noted that "[t]his will be the first real high-temperature installation in America, and its installation and operation will be followed with intense interest by all power engineers. . . . Such initiative surely deserves commendation . . . [because both] builders and buyers of this new equipment are pioneers in the development of new ranges of high station economy."[12] And for utilities such as the United Electric Light and Power Company of New York – the firm that installed two turbine units having capacities of 160 MW each in the late 1920s – *Electrical World* offered the industry's thanks for pioneering "in these attempts to raise design limits to higher and higher values."[13]

Such proud exclamations populate the trade journals in the years after World War II as well. In 1960, for example, Commonwealth Edison celebrated its purchase and operation of the first full-scale, privately financed nuclear power plant – its 200-MW Dresden plant.[14] In 1984, it continued to be proud of its technical leadership by pointing out that, next to the government-owned TVA, it constituted the largest operator of nuclear power plants.[15] Other managers took pride in their huge turbogenerators that offered the lowest capital costs or of shrewd operations and maintenance procedures combined with good technical design that permitted selling cheap electricity. TVA managers always enjoyed highlighting this last feature – offering a product that usually cost half the national average price.[16] And the American Electric Power Company still distributes a pamphlet describing its use of pioneering technologies since 1917.[17] The company hoped to add to its impressive list of accomplishments by announcing in 1988 its intention to "take a leadership role" by building the largest ever boiler that burns coal and limestone on a "pressurized fluidized bed."[18] In short, the community of utility engineer-managers valued technological efficiency for its own sake and, as one observer noted, scorned the behavior that led to simple profit maximization.[19] As demonstrated by management actions in several firms, finances were important, but not everything.[20]

This community tradition bred an unusual form of competition among some of the largest utilities. As regulated monopolies, electric utilities could not compete against others for market share within a given geographical area. But competition existed nevertheless as engineer-managers tried to surpass colleagues in other companies by operating the latest and best technologies available. In other words, "one-upsmanship" became a popular "game" in the industry. Thus, if one large company installed a new design of a 400-MW turbine generator, another might press a manufacturer for a 500-MW unit, thereby winning kudos as the firm employing the world's most powerful machine. Similar competition occurred for advanced boiler systems and generators, which meant that only rarely did two power plants in the country ever resemble each other.[21] This type of competitive environment contributed to rapid technological advances and production efficiencies in the United States as manufacturers designed new equipment. American managers took great pride in producing the cheapest electricity in the western industrialized world (except for those few countries that depended heavily on hydroelectric sources), and they attributed their success to their technological strategies.[22]

To gain the technological edge and the prestige that went along with it, utility managers sometimes asked manufacturers to develop new designs, even when engineers in the supply companies argued that the desired machines could not be produced. Samuel Insull provided an early example of this approach. In the early 1900s, he ordered large steam turbines from the General Electric Company. But when the manufacturer balked, relying on its engineering staff's assertion that 5,000 kW and larger turbines could not be built at the time, Insull threatened to go to England, where large turbines had been pioneered. General Electric eventually complied with the request and developed turbines to meet Insull's demands.[23] In the process, General Electric gained expertise and experience that helped make it the leader (in terms of market share) of turbine technology. Much later – in the 1940s and 1950s – Philip Sporn pressed manufacturers for more thermally efficient plants – and he got them from General Electric![24] And in the late 1950s, the Tennessee Valley Authority asked vendors to produce large, 500-MW units. When American manufacturers would not comply with the request, the agency bought one from the Parsons company of England. The TVA had serious problems with the machine, but the purchase had the effect of spurring design efforts by American companies toward larger units.[25]

This style of choosing and pushing technology in the United States differed from government-owned systems in Europe and elsewhere. Centralized planning characterized these systems – contrasting with the planning done by the fragmented utility industry in the United States – and this peculiar form of interutility competition did not exist. As a result, the technological style of European utility systems usually consisted of building several identical power plants, from which experience would be acquired and bugs worked out. Only then would managers move on to the next generation in technology. While American companies installed units as large as 300 MW in 1956, for example, the British nationalized Central Electricity Generating Board still commissioned units that produced less than 90 MW each.[26] This extremely conservative approach assured well-tested technology, but it also delayed benefits of new machinery. The decentralized American industry, on the other hand, sought faster advances, confident in their manufacturers' and their own abilities. When successful, the new technology operated to provide economic benefits to the companies and their customers. Managers therefore fulfilled their own psychic and professional goals while still serving their corporate and customer constituencies. No wonder utility managers enjoyed their work!

Risk sharing by leading utilities

Utility managers at leading companies did more than simply buy the most sophisticated technologies available or motivate vendors into designing advanced machines. They also took risks along with manufacturers to get the technology in the first place. Here again, consideration of the values of engineer-managers helps one understand these actions. After all, in a regulated environment, utilities obtained an almost guaranteed rate of return on their investments. They could therefore be considered passive users of technologies offered by manufacturers, who would assume all hazards of developing new forms of hardware in attempts to garner market share. Any risk taken by utilities would therefore seem to be unnecessary, for the objectives of the manufacturers alone would be enough to spur technological advances.

But some utilities offered themselves as guinea pigs nevertheless.[27] The American Electric Power Company under the presidency of Philip Sporn constituted one such company. It serves as an excellent example of an unusual form of technological leadership that resulted in benefits for the company and for the industry as a whole. As one of the few companies that emerged relatively unscathed from the holding company reorganizations after the Great Depression, the AEP (called American Gas and Electric Company until 1958) served a large geographical area that spanned from Ohio to western Virginia. Joining the company as an assistant engineer in 1920, Sporn became the firm's chief engineer in 1933 and a vice president in charge of engineering the year later. Though he acquired management power for making decisions about technologies in these capacities, he took more control when he became the firm's president in 1947. During the next fourteen years in this position, Sporn built AEP into one of the most technologically aggressive companies in the country.

During his first years as president in the immediate postwar period,

Sporn had the same concerns as most other utility managers. To meet demand and create comfortable reserve margins, he needed to build up generation capacity. Once accomplished, he then could concentrate on stimulating demand in order to attain the goals of the grow-and-build strategy. Also like most other power managers, Sporn felt an obligation to provide a service that benefited human welfare. But Sporn encouraged more than simple load growth. He persisted in working with manufacturers in order to develop the most efficient technology possible for his company's (and others') benefit. Reducing costs certainly constituted one important goal in these efforts. But as one of his colleagues pointed out, Sporn (and the people who worked with him) simply loved advanced technology. He took pride in developing and using new technology that resulted in the cheapest electricity rates in several of their operating regions, and he boasted of technological achievements.[28] For example, Sporn worked with General Electric in designing 225-MW units in the early 1950s. When the prototype tested well, Sporn ordered another half dozen.[29] Within the AEP, this approach became known as the "perils and profits of pioneering," a philosophy that meant taking on risks with new technology (and a "good deal of unpredicted trouble and expense") with the expectation that the company would enjoy the resulting benefits.[30]

This risk-taking and partnership behavior helped Sporn's AEP system win several honors for being the most technologically advanced in certain categories. In 1954, the company won the prestigious Charles A. Coffin Award (given by the General Electric Company) for using the first 345,000-volt transmission line and for developing the first very high-efficiency "supercritical" generating unit (described in Chapter 7). The utility received the award again in 1957 for the systemwide use of 345,000-volt transmission systems, for operation of 450-MW generating units, and for its leadership role in starting up eleven 215-MW units in the Ohio Valley in a thirteen-month period.[31] Other credits included being the first company (in 1950) to operate a plant with a heat rate below 10,000 Btu/kWh (above 34% thermal efficiency), the first to operate a plant using supercritical steam (in 1957), and the first (in 1960) to use a plant that broke the 9,000 Btu/kWh barrier (37.9% efficiency). Continuing to maintain the Sporn tradition after his departure as president in 1961, the AEP held several other honors, including operation of the first 765,000-volt transmission line and a series of 1,300-MW units (in 1973 through 1976) – the first run by an investor-owned utility.[32] More evidence of the firm's pride in its technological leadership can be gleaned by visiting the company's new corporate headquarters in Columbus, Ohio. Instead of obscure sculptures that often grace concrete plazas in front of huge skyscrapers, the company has installed a series of "kinetic" art pieces. One consists of the turbine rotor taken from the company's first supercritical unit.

The Duke Power Company had a similar history of technological achievement and willingness to take some risks. Serving parts of North and South

Figure 22. "We're No. 1" decal, modified by Reddy Communications, Inc. The decal displays the pride of Duke managers who operated the Belews Creek Steam Station, the country's most thermally efficient power plant. Illustration provided by Reddy Communications, Inc. Reddy Kilowatt® is a federally registered trademark and service mark and is used here with the permission of Reddy Communications, Inc.

Carolina, the utility traditionally enlisted men with engineering backgrounds as its top executives (except for it financial officers, who usually had traditional accounting experience). Never relying on outside architect-engineers for designing and constructing their systems, the company (like the AEP and a few other firms) developed a highly capable, in-house staff of engineers for these tasks. Even when the company decided to build nuclear plants, it handled the construction jobs internally, feeling that the engineering staff could do the work as well as outside contracters.[33] In many cases, the company could have been content with the standard power units that other firms used. But the strong engineering culture of Duke's leaders often led to technological strategies that incurred some risk. For example, the company employed supercritical systems in order to improve thermal efficiency at a time when other utilities waited to see whether they would prove commercially feasible.[34] The risk seemed to pay off for Duke, which still takes justifiable pride in operating the country's most thermally efficient plants (Figures 22 and 23). In short, Duke Power is another com-

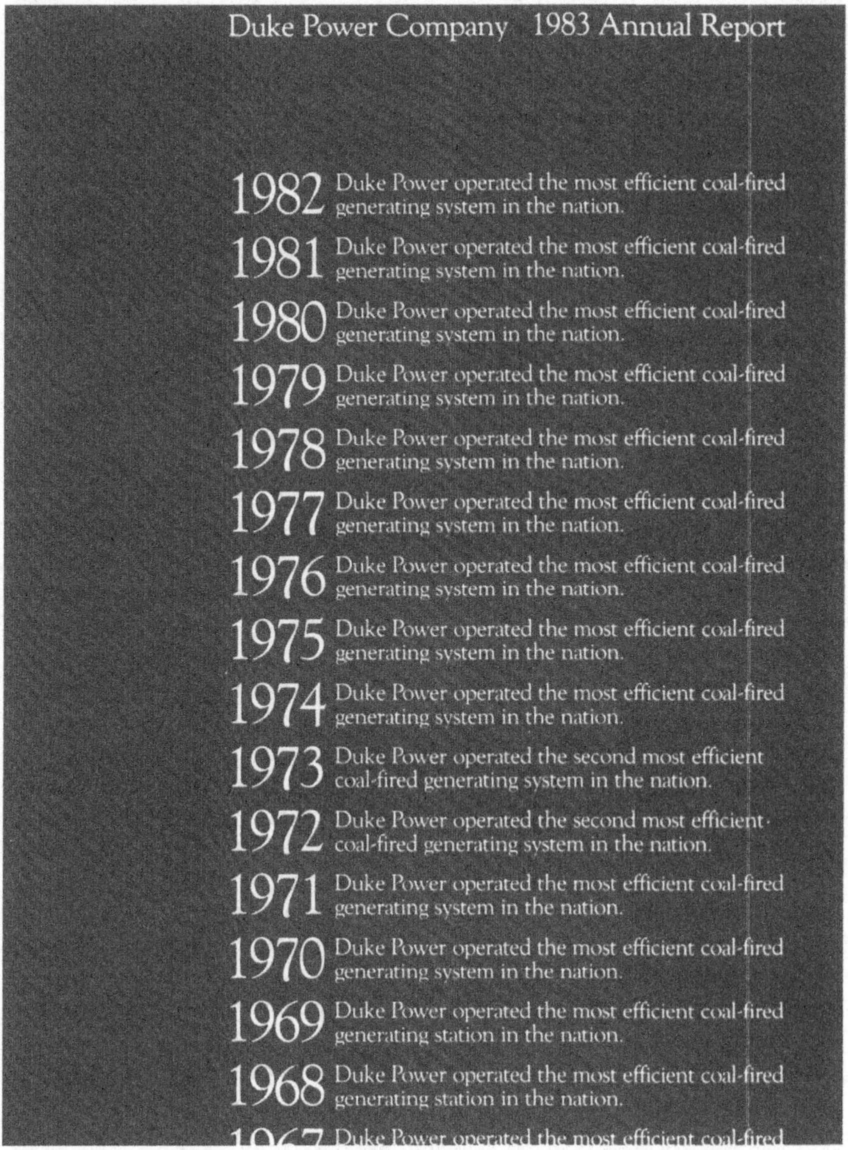

Duke Power Company 1983 Annual Report

1982 Duke Power operated the most efficient coal-fired generating system in the nation.

1981 Duke Power operated the most efficient coal-fired generating system in the nation.

1980 Duke Power operated the most efficient coal-fired generating system in the nation.

1979 Duke Power operated the most efficient coal-fired generating system in the nation.

1978 Duke Power operated the most efficient coal-fired generating system in the nation.

1977 Duke Power operated the most efficient coal-fired generating system in the nation.

1976 Duke Power operated the most efficient coal-fired generating system in the nation.

1975 Duke Power operated the most efficient coal-fired generating system in the nation.

1974 Duke Power operated the most efficient coal-fired generating system in the nation.

1973 Duke Power operated the second most efficient coal-fired generating system in the nation.

1972 Duke Power operated the second most efficient coal-fired generating system in the nation.

1971 Duke Power operated the most efficient coal-fired generating system in the nation.

1970 Duke Power operated the most efficient coal-fired generating system in the nation.

1969 Duke Power operated the most efficient coal-fired generating station in the nation.

1968 Duke Power operated the most efficient coal-fired generating station in the nation.

1967 Duke Power operated the most efficient coal-fired

Figure 23. Cover of the 1983 *Annual Report* of the Duke Power Company. The report cover lists several of the company's technological credits. Reprinted with permission from Duke Power Company.

pany whose engineer-managers willingly took steps that other utilities avoided in order to seek technological leadership.

The benefits of risk sharing included prestige and the potential for increased performance for the host utility. But it also spurred manufacturers to develop new technologies that would ultimately be used by other firms in the industry. As the AEP proudly proclaimed when winning the Coffin award in 1969 for undertaking an unprecedented program of generating capacity expansion, "[o]ur practice of pioneering in the construction of larger and larger units through the years has helped create a 'store' in which others can readily procure equipment of advanced size and design and thus share in the economies and efficiencies of scale."[35] This type of behavior by AEP and other pioneering utilities therefore benefited the entire industry, from whom they would gain still further recognition.

Of course, only a few utilities participated like AEP in advancing the state of the art. For one thing, not many managers had the temperament to be pathbreakers. Most simply allowed others to take the risks and work out the bugs in new technology. At Texas Utilities, for example, top management decided in the 1960s to help manufacturers develop technology for burning a local fuel – brown coal (lignite) – but they let other companies overcome problems in new large-scale units. Only after the technology had demonstrated reliable operation would the company purchase it.[36] A second factor that militated against too much risk taking consisted of the relatively small size of most utility systems. Only the largest utility systems demonstrating rapidly growing demand, such as AEP, TVA, Duke Power, Commonwealth Edison, Consolidated Edison, Detroit Edison, Pacific Gas and Electric, and a few others, could afford the risks involved in dealing with the newest – often biggest – technology. A small system not interconnected with other utilities, by contrast, would need to wait several years until demand caught up with the capacity offered by a big unit. A more prudent use of capital in such a case would be to build smaller units, even at a higher cost per kilowatt, since the equipment would find use quickly and begin to produce income immediately. In other words, absolute growth in capacity requirements was important for economies of scale to be realized and for big technologies to be used.[37]

The apparent precedence of engineering values over "normal" management behavior in the electric utility industry has been explained partially by invoking the role of a community of tradition. But why did this tradition remain successful in this industry while being blended into the world of corporate capitalism in others? For one thing, regulation may have played an important role. Because utilities generally could count on a reasonable rate of return from their investments, managers felt less concerned than their unregulated counterparts about running the most economically efficient firms. They did not require the same rigorous attitude toward profit maximization that characterized unregulated firms, partly because they knew that if their profits became too large, they would be required to make refunds to customers. As long as the companies provided reliable service and reasonable (i.e., declining) rates, they would be protected from too much scrutiny and would be able to do pretty much what they wanted. This meant making sure their companies had sound economic bases, but it also meant that managers could pursue goals as engineers who enjoyed their technologies and who saw themselves as makers of a better world. In such an environment, utility managers contributed to progress in power equipment.

In what is known as the "Averch-Johnson effect," regulation may also have encouraged engineer-managers to choose large-scale, capital-intensive technology. Named after the economists who first described it, the effect occurs because state regulators generally permit utility companies to earn a return only on capital expenditures.[38] In contrast, expenses for fuel and labor are passed on directly to consumers without profit. Regulation therefore promotes the expansion of the rate base – the sum of all capital investments (less depreciation) that earn a return – through the use of capital-intensive technologies. Hence, according to this account, regulation fosters the purchase of large-scale technologies that substitute capital for labor and fuel. The Averch-Johnson effect therefore reinforced managers' "natural" desire to employ advanced equipment, because purchases of big technology constituted a rational strategy given the regulatory constraints.[39]

Meanwhile, the behavior of utility managers did not necessarily conflict with standard business practices of managers in other industrial enterprises. As argued by Robert Reich, American managers in the period between 1920 and 1970 became obsessed with huge size and mass production. They developed organizing principles that gave central importance to the use of large-scale, high-volume production technologies featuring long runs of standardized products. In the automobile, steel, rubber, and chemicals industries, for example, these technologies generated vast economies of scale, tremendous productivity improvements, and reduced unit costs.[40] Utility engineer-managers, therefore, did not hold a monopoly on their fascination with large scale and advancing technologies.[41]

Managers of several utilities, then, played major roles in the development of advanced forms of technology throughout the twentieth century. They encouraged manufacturers to build more elegant technology so they could get credit for using the "best" and most exciting machines, and they took on extra risks to obtain the prestige. Sometimes the risks to gain technological supremacy meant an economic trade-off because of the increased complexity and difficulty in operating high-efficiency units. As one industry insider noted, "the most efficient plant is rarely the most economic . . . [and] is, in economic terms, rarely worth building."[42] Nevertheless, the quest for technical prestige rarely caused much trouble in the 1950s and 1960s.[43] Moreover, the extra risks were a price readily paid by managers who retained their engineering values and goals even as they became leaders of large business enterprises.

6 The mid-1960s: At the pinnacle of success

The rising share prices of utility stocks through the mid-1960s reflected the enthusiasm and sense of success felt by people in the utility industry. As the utility averages doubled between 1958 and 1965[1] – soaring to heights not seen since before the Depression – managers could feel a justifiable pride in their values as engineers and their successful stewardship of technological systems that appeared to bring widespread benefits. To be sure, the electrical equipment manufacturers contributed to the development of the technological advances that made this success possible, but the utility engineer-managers provided the market and the incentive for the technology, and they were the ones who dealt directly with the public in providing a valuable service.

A few sets of data will help in summarizing the successes made by managers in the power industry. As noted in this book's introduction and displayed in earlier figures, rapid growth in electricity usage became the norm. From 1900 to 1920, usage galloped ahead at a 12% annual growth rate. From 1920 until 1973 (with only one interruption during the Great Depression), the growth rate advanced on average at a 7% annual pace. The growth becomes more impressive when compared to other energy sources: electricity consumption increased 4 to 5.5 times faster than the rate for all energy sources together[2] and more than twice the 3% annual growth rate for the economy in general.[3] Obviously, electricity had merits in terms of versatility and productivity enhancements that made it the energy source of choice.

Electricity became popular partly because it enhanced the productivity of the industries and individuals who used it. At the same time, the power industry improved its own productivity through the use of incremental advances in technology. For example, the thermal efficiency of power technology increased on average from about 4.5% in 1907 to 33% in 1962,[4] with the best units reaching as high as 40%. These improvements, dependent largely on increased steam temperatures and pressures, enabled utilities to produce each kilowatt-hour with less fuel. Growth in the scale of power units, rising from 5 MW in 1903 to 1,000 MW in 1965, provided economies that also helped the industry make more efficient (and productive) use of its resources, with the largest contribution coming from the increased output rating of turbine generators.[5] Together with associated improvements in transmission, distribution, and system controls, these technical advances helped elevate the industry's productivity to levels that could not be reached by any other industry. Improving an average 5.5% per year from 1899 to 1953, electric utilities' total factor productivity dwarfed that of the entire private domestic economy, which only showed a 1.7% per year improvement for the same period. High productivity growth rates continued for the utility industry into the 1960s.[6]

Largely as a consequence of these productivity improvements, the unit cost of electricity declined in a way that benefited utility companies, investors, and customers alike. As Insull discovered, companies could take advantage of new technology and increasing consumption to reduce costs and pass along some of the savings to consumers: in 1892, residents paid about 92 cents per kWh (in adjusted 1967 terms). By 1967, a residential customer paid only 2 cents for the same amount of electricity. (In adjusted 1986 terms, the respective rates were 316 and 7 cents per kWh.) Duke Power Company accurately assessed the cause of this trend in a 1960 advertising that proclaimed "[y]our greater usage plus technological advances are responsible for the lower cost per kilowatt-hour."[7]

These average figures for the entire industry do not do justice to the larger, more aggressive utility firms that exploited the grow-and-build strategy to the hilt. One such company was the Virginia Electric and Power Company, the firm that proudly plastered the word "GROWTH" on its 1968 annual report cover. As the following data indicate, the book's cover did indeed tell the inside story. Between 1951 and 1967, the company installed facilities that increased its power production capability by 432%. At the same time, the output of electrical energy jumped 341%. But because it used new and more efficient technology, the company improved its system's thermal efficiency by 26.5%, contributing to lower costs. Combined with economies of scale, reduced costs translated to lower prices for consumers – down 23% in the period for residential customers and 14% for industrial users.[8] And just as the growth strategy promised, the company's financial picture brightened. Even with decreased consumer prices, the amount of revenue from each customer increased by 118%. Stockholders exulted, too, as they watched the company's earnings per share increase by 283% and share price rise by 533%.[9]

Basically, two groups of participants effected these tremendous technological and financial gains. The first consisted of the manufacturing community that designed new equipment. In the turbine-generator business, the vendors first used a design-by-experience approach and then, beginning with the rush of orders after World War II, a design-by-extrapolation method. Both approaches appeared to work well, producing a continuous array of better technology offerings. And there appeared to be no reason to believe that the future would bring anything but continued technological success. In 1964, when the Federal Power Commission issued a report on the status of the electric utility industry, people felt secure that the trends of incremental progress in turbines, generators, and boilers would remain intact.[10] For example, the report confidently forecast that turbine units would reach 1,500 MW by 1980.[11] Other industry observers expected even more powerful machines.[12] Enthusiastic about the future, the FPC report predicted that the trend toward technological improvements, increased usage, and lower costs would *accelerate*. "The electric utility industry of the United States stands at a threshold of a new era of low cost power for all sections of our country," the report indicated. "Many exciting new technological developments point the way. Larger and larger machines are being built which can generate electricity at progressively lower costs."[13] It seemed that nothing could impede this progress.

Utility companies constituted the second participant, benefitting directly from the new technology and their excellent relationship with manufacturers. Creating a large market for advanced equipment and sharing in some of the risks involved in developing it, utilities trusted their suppliers to provide the ever-progressing technology that was the necessary prerequisite for the success of the grow-and-build strategy. Happily for all parties, no one was disappointed. Utilities promoted electrical usage aggressively in the 1960s, using privately developed campaigns or adopting the "Live Better Electrically," "Gold Medallion Home," and Reddy Kilowatt programs.

But utility managers did not just content themselves with the economic "bottom line." Since the beginning of the century, they had developed a value and belief system around technological progress. Engineers generally had great affection for their technology and enjoyed using advanced versions of it. They also felt proud that their skills in developing and using sophisticated equipment brought benefits to the companies' customers and shareholders alike. Therefore, if the first eighty years of the power industry's history left a legacy that would influence later actions, it was not just a history of success in employing advanced technologies. It was also a sense of accomplishment derived from knowing how to use that technology to benefit all elements of American society.

During the period of rapid expansion and success, a third participant played a largely invisible and benign role supporting utility companies' activities. The participant was the state regulatory body. Though every state's regulatory body had different origins, they generally shared several traits, including the selection of commissioners by political office holders or by the public at large through elections. In 1979, for example, voters elected utility commissioners directly in only eleven states. In the remaining states and the District of Columbia, elected officials appointed them.[14] Once in their positions, commissioners performed two tasks. First, they oversaw a legalized "natural" monopoly in such a way that it acted as if a competitive free market existed. In other words, the regulatory body protected the public from abusive monopoly practices while also assuring reasonably priced and reliable service from the utility company. While safeguarding the public, commissioners also looked after the interests of utility companies by guaranteeing their financial integrity. Because utilities served only specific regions and could not easily move their businesses, they naturally desired assurances that they would remain viable and be able to attract investment capital.

This two-faceted responsibility of the public utility commissions may appear to have been contradictory. Yet, few conflicts arose between utilities and their customers because companies used new and more efficient technologies, thereby continually reducing their marginal costs of producing power. Because utilities passed along some of these cost savings to consumers, few people complained about a service whose costs countered the general trend toward cost of living increases. As a result, regulatory actions tended to reinforce the grow-and-build strategy.[15] Representing (in part) another group of participants – electricity users – regulatory bodies therefore had a fairly easy task. As one regulatory official reminisced in 1981, "the task of state regulators was essentially one of distributing among rate payers the benefits of progressively higher efficiencies achieved by utility managers – not bad work if you can get it."[16] And, as one academic observer of electric utility regulation noted, regulators had a "happy task of watching the industry become more efficient and, on occasion, of negotiating rate reductions."[17] He also concluded that as "long as technological opportunities promised lower costs through increased usage, utilities were given a free hand to adopt rate structures that promoted consumption. Large users clearly benefited from this policy, but so did consumers, investors, and, as a result, regulators."[18] Undoubtedly, utilities did not realize all possible economies, and critics have pointed out that the industry often failed to achieve maximum economic efficiency.[19] Yet, as long as costs declined while utilities continued to supply electricity with ease, they received good treatment from state regulatory bodies. This happy situation by which utilities won monopoly status in return for providing cheap and abundant supplies of electricity has been called a "social contract" by some utility managers.[20] It was a contract that served the industry well for more than half a century.[21]

Finally, the investment community constituted another – almost silent, but vitally important – stakeholder in the electric power matrix. Made up of brokerage firms, investment bankers, and individual investors, this group provided the capital needed for an industry that expanded faster and more consistently than the general economy. Utility stocks generally provided secure and growing dividends along with rising share prices, offering a safe investment to retired people and others who sought growth, income, and capital preservation. Though investors might have been able to make money faster with speculative stocks, few ever lost money by holding utility stocks over the long term. Investors simply viewed the utility industry after World War II as a safe, consistent – though unexciting – place to put one's nestegg.

To an observer in the mid-1960s, then, the utility industry appeared to be a picture of success. Manufacturers produced consistently improving technology, and utility companies employed it to benefit their companies, customers, and investors. Moreover, utility managers had incorporated many of the assumptions and standard business practices for encouraging growth and technological progress into their management culture, insuring that the goal of mutual profit for all participants in the electric power matrix would be maintained. Though not articulated at the time, these stakeholders in the electric power matrix had forged an implicit consensus concerning the choice, management, and regulation of a technological system. As long as benefits continued to accrue to everyone, the consensus would remain intact.

Part II
Stasis

"We hoped the new machines would run just like the old ones we're familiar with, and they sure as hell don't."

– Jack Busby, president of Pennsylvania Power and Light Company, commenting on experiences with "advanced" power technology in the late 1960s.

7

Technical limits to progress in the 1960s and 1970s

At the same time as utility managers revelled in their success, the technological basis for their celebration had begun to erode. Instead of improving beyond the 40% mark, thermal efficiencies for the best plants plateaued in the 1960s. And after accelerating since World War II, the capacities for the most powerful units apparently reached an impenetrable barrier in the early 1970s. Here, then, were the manifestations and outward signs of technological stasis.

Thermal-efficiency plateaus

Limits to thermal-efficiency improvements occurred after substantial gains in the post-World War II period. From 1947, when power plants transformed only 21.8% of a fuel's energy content into electricity, the average thermal efficiency rose to 32.9% in 1965.[1] Several factors accounted for this improvement. Many relate to advances in metallurgical knowledge gained during the war and used in aircraft and artillery. After the conflict, when power-generation equipment could be built again, engineers in the industry took advantage of new materials and knowledge. Newly developed "superalloy" steels that resisted metal fatigue and cracking, for example, allowed engineers to design larger components for more power output. At the same time, the new metals withstood the effects of using steam at higher temperatures and pressures, thus providing greater thermal efficiencies.[2]

Though temperature ratings did not increase as dramatically in the immediate postwar period as in the 1920s and 1930s, thermal frontiers advanced again in the late 1950s, when manufacturers announced the development of supercritical boilers. These units produced steam at pressures in excess of 3,206 psi and at temperatures above 1,050 °F. Under these conditions, water turned into dry (unsaturated) steam without boiling, thereby saving the latent energy needed for the conversion.[3] The first such supercritical unit began service in 1957 at the Philo Plant of the Ohio Power Company (part of the American Gas and Electric system). Operating at a pressure of 4,500 pounds per square inch and at a temperature of 1,150 °F, the plant established a record for the most thermally efficient unit of its time, designed to produce power at an efficiency of just over 40%.[4] Though the plant's components required special alloys, engineers and utility managers believed the benefits of supercritical systems would offset the extra cost of the metals. A similar view emerged for subcritical units too. In other words, utility and manufacturing managers felt that all new machines (supercritical and subcritical) could benefit from using higher temperatures and pressures. The conviction led to vendors producing prototype units whose steam temperatures crept up at a rate of about 12 °F per year, from about 900 °F in 1945 to 1,106 °F in the 1950s (for GE units). At the same time, thermal efficiency improved as steam pressures increased, from about 2,500 psi (for the maximum pressure used) in the 1940s to greater than 4,000 psi in 1955 for the most advanced new units[5] (Figure 24). (Also see Appendix B for other advances in thermal efficiency.)

Efficiency improvements that reduced fuel consumption per kilowatt-hour by 25% in the twenty years after World War II signified tremendous economic and technical progress. Unhappily for the industry, however, continued advances in the 1960s simply did not occur. Fluctuating within a narrow range after the mid-1960s, the average efficiency rate for all plants in 1980 registered only 32.5%, a figure slightly lower than 1965's rate.[6] While disturbing, this flattening trend occurred at a time when nuclear power plants began contributing a growing proportion of electricity. Because nuclear plants produce steam at lower temperatures and pressures than do fossil plants, they exhibit inherently lower thermal efficiency. When data from nuclear plants are lumped together with those from fossil plants, therefore, the average thermal efficiency naturally declines. True as this is, the average efficiency for fossil-fueled steam electric plants alone – that is, after excluding nuclear power plants – showed the same flattening trend. Rising from 21.7% in 1948, average efficiency reached 32.7% in 1965. For the next fifteen years, essentially no progress occurred. In 1980, average efficiency rested at 32.8%.[7] When looking at the *best newly in-*

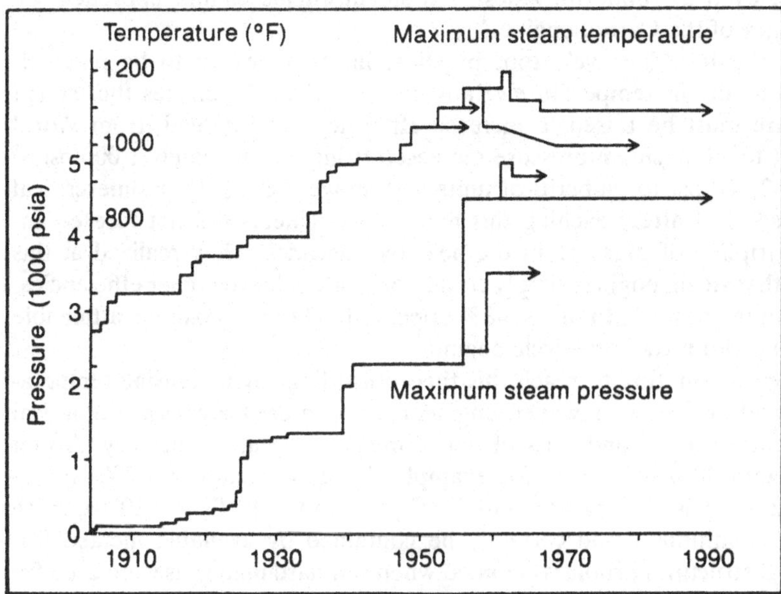

Figure 24. Graph illustrating how maximum steam pressures and temperatures of turbines increased until the early 1960s, yielding greater thermal efficiencies. Reprinted with permission from Vol. 12, *EPRI Journal* (December 1987), p. 12.

stalled fossil plants in the country rather than the average of all plants, one still sees a similar pattern. These plants represented the state of the art at the time, but even they showed slipping performance. After a supercritical unit reached about 40% in 1961, the thermal efficiency for the best new plant in 1977 declined to 36.0%.[8]

For fossil units, the thermal-efficiency plateau resulted from theoretical, economic, and metallurgical causes. Perhaps the easiest to understand dealt with thermodynamic theory. While steam systems appeared to offer much room for gain – from 40% or so to the ideal of 100% – this was not the case. After all, thermal efficiency depends on the temperatures and pressures of steam as well as the temperature of the condenser. Even when using the Carnot cycle, which characterizes an ideal engine that operates on heat energy, inherent limits with available metals (i.e., they cannot be heated to infinite temperatures) and a realistic temperature of condenser water (above the freezing point) would establish a maximum efficiency of less than 75%. (See Appendix B.) In reality, however, stricter limits apply because the Carnot cycle does not represent a true steam-turbine system. Unlike an ideal gas, steam evaporates and condenses within the range of temperatures used in engines and is therefore governed by Rankine-cycle theory, which dictates a lower efficiency than Carnot theory. In other words, even a wonderfully designed steam engine cannot approach the efficiency of the Carnot engine.[9]

For the Rankine cycle too, practical limits appeared to be reached. Given a certain temperature, Rankine-cycle theory indicates that steam pressure must be raised to increase efficiency. And indeed, from World War II to 1962, steam pressure increased from approximately 1,000 psi to about 2,400 psi for subcritical units and above 3,206 psi for supercritical boilers.[10] But after reaching this milepost, engineers did not foresee another tripling of pressure in the next two decades. They realized at this point that steam engines simply could not easily achieve higher efficiencies. Even in the best of situations, 48% appeared to be the maximum attainable efficiency for a Rankine-cycle engine.[11]

When attempting to reach this theoretical limit by increasing temperatures and pressures, however, engineers encountered severe metallurgical problems – the second class of impediments to greater efficiency. A unit can attain 40% efficiency, for example, by using steam at 1,200 °F (reheated to 1,050 °F) and at about 3,200 psi.[12] In the 1960s and 1970s, these steam conditions could certainly be contained by available metals. But unusual structural problems loomed when standard materials were used for long periods – years or decades – in turbine and boiler equipment. For example, ordinary alloy steels eventually deform (or "creep") in steam environments of temperatures above 1,200 °F. Other problems, such as "stress corrosion cracking" in turbine blades and piping also occurred under the combination of high mechanical stress and a corrosive environment of steam at high temperatures and pressures.[13]

Many of these metallurgical problems showed up in the boiler tubes used to transport water and steam from the furnace. Statistics on boiler-tube failures during the ten-year period from 1964 to 1973, for example, indicated that the forced downtime caused by these failures exceeded the sum of deficiencies from all other components.[14] Problems with the metals in the tubes, combined with contaminants in water appeared to be major culprits in causing failures. As operating engineers "cycled" power units from full capacity to lower capacity (and the reverse) in order to meet changing load demands, tubes underwent temperature variations that caused fatigue and metal creep. Then, as tubes cooled, scale fell off their inside surfaces, causing corrosion and erosion damage to turbine nozzles and other components.[15] Cycling of plants also increased opportunities for the shafts and casings of turbine-generator sets to crack due to thermal stresses.[16]

Some of these problems had existed for several years and with older units. However, in boilers that employed steam at lower temperatures and pressures, many of these problems did not occur with such annoying regularity. Control of water chemistry, for example, was not as critical at lower temperatures, largely due to decreased reactivity of most chemicals. According to a rule of thumb, chemical reaction rates double as temperatures increase by 10 degrees Celsius. Therefore, an increase of just 100 °F (from say 1,000 °F in a 1950s-era unit to 1,100 °F in a unit installed in the 1960s – a rise of 56 °C) increased corrosive activity fiftyfold.[17] While research continued throughout the 1970s and 1980s, the problem with boiler tubes remained. "It is a measure of the enormity of the problem that such problems, though long known to boiler designers and operators, still plague the utility industry," declared one researcher in 1983.[18]

When reaching for higher temperatures and pressures, then, engineers needed to use special, "high" alloy metals that had greater resistance to increased stress. Known as "austenitic" steels, the metals contained elements such as nickel, chromium, molybdenum, and manganese mixed with iron and carbon. The small jump in temperature and pressure became expensive, however, because austenitic steels cost three to four times that of the conventional carbon steels.[19] As a result of these considerations, the marginal cost of improving thermal efficiency became very expensive. In 1957, engineers calculated that the cost of improving efficiency from 38 to 39% was about $3 per kilowatt. An increase from 39 to 40% cost about $5 per kilowatt. But the next one-percent increment would cost about $12 per kilowatt. The great addition to capital cost suggested to many engineers that higher thermal efficiency simply did not make good economic sense, especially in the 1950s and early 1960s when fuel costs remained low.[20]

But even when utilities built plants using components having these expensive steels, metallurgical problems often followed. Harder to machine and work with in large sizes, these metals also required totally new learning experiences, since they often acted in unexpected ways.[21] As one utility metallurgist put it squarely in 1982, "we don't know how to handle high strength materials – at least not on a commercial basis." In aircraft that received regular and frequent maintenance, these materials had become commonplace, he noted. But "when we use high strength material in the power plant, we suddenly are susceptible to new diseases like stress corrosion cracking, which was literally unheard of with low strength materials."[22]

These metallurgical problems showed up in a lower "availability factor," a measure of the amount of time a unit produces power. New forms of technology traditionally demonstrate lower availability factors as operators detect "bugs" and learn from experience how to maintain units. Manufacturers traditionally depended on this learning to design the next version of units that avoids similar problems. Nevertheless, engineers began noticing as early as the 1950s that steam plants demonstrated poor availability factors when employing higher temperatures and pressures. Plants that operated at temperatures of between 1,040 to 1,060 °F from 1955 through 1961 had availability factors of at most 88%, whereas units using steam at between 900 and 955 °F were available for 91% of the time.[23] Engineers also found a similar correlation for units that employed high-pressure steam: the higher the pressure, the lower the availability.

These problems with pushing back the thermal-efficiency frontier motivated a change in policy among utilities in the 1960s. As noted in *Electrical World* in early 1960, two pioneering utilities (Philip Sporn's American Electric Power Company and the Public Service Electric and Gas Company) had announced their retreat from 1,100 °F steam to about 1,000 °F. One of the editorialists for the widely read industry magazine regretted the "dramatic withdrawal from use of austenitic steel at 1,100 °F and above." He hoped that the retreat would stimulate research that would lead to a breakthrough in metallurgical science. Such an effort, he noted, "is necessary to justify not only an early return to 1,100 °F but also an advance in temperature that will bring compensating thermal gains."[24] But when *Electrical World* surveyed the country's power plants in 1964, it found that the trend away from high temperatures had continued, with pressures decreasing as well. While the mean steam temperature declined six degrees from 1,009 °F since 1962, the mean pressure for 1964 stations fell 92 psi to 1,846 psi. Once again, an *Electrical World* editorialist noted that it "is certainly clear that metallurgy today remains the barrier to further gains in the station heat rate. And, until steam temperatures can be turned upward again, we can anticipate only modest thermal gains from larger units and refinements in the steam cycle."[25]

Eight years later – in 1972 – the situation did not look much brighter. An *ad hoc* committee on energy research and development, convened by the National Association of Regulatory Utility Commissioners reported unhappily:

> The primary reason for the leveling-off in efficiency of the steam cycle is the inability to improve metallurgy in the boilers and turbines to withstand temperatures substantially above 1,050 degrees Fahrenheit. The industry has been compelled to retreat from its efforts to raise boiler temperatures above 1,050 or 1,100 °F. Here is a field where one would think that scientists could make a large contribution.[26]

And as summarized by a TVA manager who supervised the introduction of new technology from 1944 through 1970, the situation was one in which manufacturers "had to find materials that would stand stresses of higher pressure and temperature development. Essentially what happened was that technology ran up against the failure of metallurgists to find something that would go further."[27]

Unit-size barriers

Technological stasis also became evident in the growth of the scale of power equipment. After rising during the 1960s, unit capacity stabilized in the early 1970s. In 1973, for example, 1,300 MW marked the output of the most powerful new unit. The following two years saw new additions at that level – fossil units installed by the TVA and American Electric Power – but through the 1970s, new turbine generators generally carried nameplate ratings of less than 1,000 MW. Meanwhile, average sizes plateaued as well.

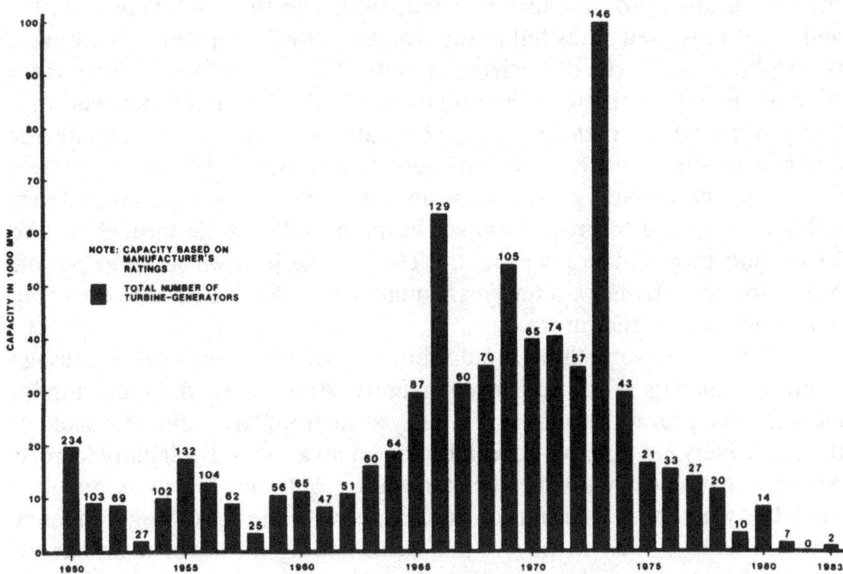

Figure 25. Capacity and number of turbine generators ordered in the United States. An ordering "panic" hit the utility industry in the late 1960s and peaked in 1973. Reprinted with permission from the American Electric Power Company.

From 1960, when the average size of both fossil and nuclear units stood at about 240 MW, nuclear units soared to an average of about 1,000 MW in 1975 and grew only slightly – to 1,100 MW – by 1980. The large size reflected the belief that nuclear plants took better advantage of scale economies than fossil plants. Average sizes of fossil plants rose to about 600 MW in 1970 and remained there into the 1980s.[28]

The causes of problems in increasing the sizes of units in the 1970s stem from technical and business management concerns. Some reasons appear extremely logical and simple. Others (such as availability problems) are still not widely acknowledged or have causes that cannot be pinned down. The most frequently advanced reason for the plateau and decline in unit size concerns the utility industry's inability to project growth rates of electricity usage after the 1973 oil embargo. With growth rates fluctuating annually (though in a general downward trend), utility managers realized that large-capacity additions would not be used when their companies completed building the plants. Lack of use translated to misallocation of economic resources, so managers either delayed new construction or ordered smaller increments of capacity.

This argument appears to have some validity. Due to the dramatic decline in usage during 1973, orders for new power units of all sizes dropped significantly – from 146 units in 1973 to only 43 in 1974 (Figure 25). And the size of units ordered also declined, so that in the years between 1972 and 1982, most new units fell in the 400- to 800-MW category.[29] However, the explanation for the downsizing of units based on declining growth rates of usage is not complete. While managers in the late 1980s believed that low growth rates (in the range of 2 to 3% per year) would remain for another decade or more, most managers did not expect slow growth to be a fixture on the electric power scene in the 1970s. As will be noted later, utilities continued to predict growth in the 6 to 7% range throughout the 1970s, and they had to plan for it.[30] Thus, while it made sense to put off orders for new plants for a few years, managers still felt a need to order big units sometime in the future.

Perhaps, then, something in addition to reduced growth rates in usage motivated managers to order smaller units. And maybe the more fundamental reason for utility managers to shy away from large units consisted of their relatively low reliability, which showed up as poor availability factors. Already in the early 1960s, the correlation between big units and low availability had become noticed, though not with the disturbing frequency it would later. Utility managers had been hoping for lower maintenance costs with more powerful units, but they discovered that they had bought equipment that demonstrated lower availability instead, requiring more labor and time to bring them back into service. As members of the committee that wrote part of the 1964 *National Power Survey* noted, operating experience suffered as much as 3 to 4% with large units. Units smaller than 200 MW had availability factors of about 90%, but those rated between 200 and 325 MW demonstrated availabilities of about 86% between 1955 and 1961.[31] Data collected between 1960 and 1972 indicated that fossil-fueled units in the United States had a forced outage rate – the amount of time during a year that a unit cannot operate because of failures – as much as five times greater for units larger than 600 MW than for units in the 100-MW range.[32] Later data showed the same effect: between 1967 and 1976, units larger than 800 MW had been forced out of service more than 16% of the time. More modestly sized machines, those between 200 and 300 MW, had forced outage rates less than 6%.[33]

The simplest explanation for the correlation between large units and low availability concerns sheer size and complexity of large plants. Whether using nuclear or fossil fuel, these plants required congeries of redundant systems to ensure safety and good performance. They also needed massive pipes, fittings, and connectors – in general, large components and many more of them than a smaller plant. Even if a unit operated perfectly, it would require periodic maintenance and overhaul, forcing its removal from the power grid so that all movement stopped. Unless utilities hired more laborers to maintain the system – thereby losing some of the advantages of large unit size – extra time would be spent on servicing the machine.[34] Moreover, the large masses took time to cool down – and then to reheat when operation began again – and this time relates proportionally to the mass of the components. Maintenance personnel could not work on hot components, so the time to cool down and reheat constituted essentially "dead," unproductive time that reduced the availability factor.[35]

Supercritical units proved even more complex and "unforgiving." Interviews of plant managers by Robert Gordon, an economist at Northwestern University, revealed a general condemnation of the "advanced technology" that offered high thermal efficiencies. Some managers reported problems with "complex valving" and an associated "burden of maintenance." One manager complained that the units had been designed in the laboratory with little feedback from the "real world." Others commented on high forced-outage rates and metallurgical problems that showed up as gas leaks in furnaces.[36] And while they demonstrated great thermal efficiencies, supercritical units also required lengthy and carefully timed start-ups and cooldowns (which reduced the amount of time they operated at optimum levels). The Eddystone 1 supercritical unit provides an example of these maintenance and reliability problems. After setting a new thermal efficiency record in 1962, the unit spent most of 1963 being repaired.[37]

Similar difficulties in operating and maintaining other supercritical units contributed to their declining popularity in the early 1970s. After reaching a high point of 62% of new capacity additions in the 1970–4 period, the installation rate of supercritical units dropped to 31% in the next five-year period and just 6% in 1981 and 1982.[38] This assault on and retreat from a technological frontier appeared unprecedented, an event that some observers considered "unique in the history of technological advance in the electric power industry."[39]

Nuclear plants also became an overwhelmingly complex form of technology as a result of imposed safety regulations and management problems. Exemplifying the lack of simplicity in equipment and management are the 90,000 drawings needed to describe a typical unit, or the 8 million sheets of paper required to support its quality-assurance program. Duke Power Company, builder of three nuclear plants, came up with another indicator of complexity: the number of parking spots at a nuclear station. When designing its Oconee plant, managers thought they could operate it with about 100 people. Because of regulations that required extra control systems and security provisions, however, the station in 1986 employed more than 1,000 people, requiring construction of several new parking lots![40] And as the manager of TVA's Engineering and Construction Department put it, building a nuclear plant was "like building a watch." Designers had to be "exact and precise," but they still made mistakes.[41] A recent study performed by the utility industry's Electric Power Research Institute reinforces these views, noting that complexity has increased "greatly out of proportion to size" resulting, to a large extent "from the addition of equipment as retrofits in response to changing regulations, rather than as part of a plant's integrated design." A large, 1,000-MW reactor, the report indicated, could include as many as 40,000 valves, many of which would have been unnecessary if the plant had been designed properly from the start with the new regulations in mind.[42] Operation of such a plant would *a priori* be more difficult than a smaller one designed on older and less complex principles.[43]

Simple physics also played a role in establishing size limits. In turbines, for example, long blades wrench power from low-pressure steam at the end of the machine. Because the shaft rotates at 1,800 rpm for turbines in nuclear plants and 3,600 rpm for fossil units, tremendous centrifugal forces build up at the blades' tips and "roots," which hold the blades to the turbine shaft. (A 44-inch last-row blade travels about 950 miles per hour, creating an outward pull of 33 tons on a 7-pound blade.[44]) Erosion and corrosion of the blades, along with unanticipated aerodynamic forces acting at the tips, caused severe problems for at least one manufacturer, Westinghouse.[45] Stress-corrosion cracking and other problems also occurred at the blade's base. In short, there appeared to be some physical limits to the size of last-row, low-pressure turbine blades due to metallurgical problems.

Generators faced some limits as sizes increased too. Through advances in hydrogen and water cooling, these machines could produce large amounts of power in housings that remained relatively small – certainly just small enough for efficient transportation on railroad cars. But as ratings increased above the 1,000-MW milestone, designers realized that water-cooling techniques became stretched. The realization spurred General Electric to begin work on the next type of cooling system that used gas at about 4 degrees *Kelvin* – near absolute zero. Ultimately successful in the early 1980s, managers nevertheless shelved the 20-MW prototype because of a lack of interest in larger units.[46]

Environmental concerns appeared to set some economic limits to size, too. As was discovered in the early 1960s, thermal efficiency reached an effective barrier due to metallurgical problems that resisted efforts to raise steam temperatures and pressures. To reduce costs, then, utilities sought "economy of scale with larger and larger generating units."[47] But bigger units created more gaseous waste and thermal pollution and taxed the local environment more than smaller units, requiring construction of extensive pollution abatement facilities. In the past, when managers could essentially disregard a large part of the environmental costs of producing power (due to less stringent laws), utilities saw great advantage in building huge units. But when they included the expense of pollution-control equipment in the cost of power stations, the economic incentive for constructing bigger generating units no longer appeared to exist.[48] The conclusion from a study of these countervailing trends did not console utility managers: "[t]he drive for lower energy costs and better generating station efficiency today is on a collision course," noted *Electrical World* in 1967.[49]

Several of these technical barriers occurred in all large steam plants, but others existed only in large nuclear units. Steam exposed to radiation in nuclear "boilers," for example, contains elements such as oxygen that become transformed chemically. These "radiolytic" products in the steam corrode the pipes that bring steam to heat-transfer mechanisms, which provide heat to "clean" steam that passes through the turbines. But even the clean steam gets contaminated with halogens, which attack the bases of turbine blades through stress-corrosion cracking.[50] The contaminated steam also affects pipes and valves, causing availability rates for nuclear plants to be generally lower than those for fossil units.[51] The severity of these problems spurred the Nuclear Regulatory Commission to establish a Pipe Crack Study Group in 1975.[52]

Besides these problems, which have origins in clearly understandable technical effects, others appear to defy complete explanation. Part of the reason for the unknowns has to do with the fact that utilities choose from a large selection of equipment made by different manufacturers and installed by several construction contractors. Moreover, manufacturers design equipment to meet the specific operating requirements of utilities that use special fuels or operate in especially hot or humid environments. In building a nuclear power plant, for example, utilities could choose from one of the fourteen largest architect-engineering firms, which could select boilers from one of four vendors. These choices produce fifty-six possible combinations for builders and boilers.[53] When considering turbine-generator units and auxiliary units, the possible permutations increase still more. In short, rarely are two power plants exactly alike. Therefore, problems in piping, which may cause low availability in one unit, may not show up in others, where turbine-blade defects occur. Because of the inability to duplicate these problems, manufacturers sometimes attributed the causes to poor utility maintenance rather than to anything fundamental in the hardware itself.

Metallurgical weakness at high temperatures and pressures, unreliability of large plants, complexity, simple physics, and environmental concerns – these constituted some of the culprits that presented barriers to improvement in electric power technology. Coming quickly in the 1960s and 1970s after years of success in improvement, these barriers disappointed engineers and utility managers. Put simply by one engineer who has tried to examine the phenomenon of limits, the technology had simply "hit the wall."[54] Or, as economist Robert Gordon concluded, "[p]lant designers appear to have run into partly *unanticipated technical barriers* that caused them to build plants that were too large, too complex, and which required a high and unanticipated level of maintenance expenditures."[55]

8 Design deficiencies and faulty technology

The apparent limits to progress in electric power technology cannot be attributed totally to "laws of nature" or metallurgical failures. While the Carnot and Rankine cycles dictated diminishing returns to improving thermal efficiency, manufacturing firms contributed to equipment problems and unreliability of large units by improperly applying the design-by-extrapolation technique. They used the technique in response to heavy demand for larger units, and it appeared to work well for a time. In the mid- to late 1960s, however, it failed, resulting in a "crop" of unreliable large units. Stasis, in this case, occurred partly because managers reacted improperly (as we see in retrospect) to market pressures and used a design philosophy that promised more than it could deliver.

Problems with design techniques

Manufacturers felt intensified pressure from utilities in the 1960s for an increasing amount of new and larger-scale equipment. When the entire utility industry (public and private) had reached the level of about 236,000 MW of installed capacity in 1965, for example, annual orders should have been placed for about 20,000 MW to accommodate demand growth by the time the plants would operate a few years later.[1] But after a glut of ordering in 1966 – exceeding 60,000 MW for 129 units – orders still outpaced demand growth by remaining in the 30,000-to-60,000 MW range from 1967 through 1972 (Figure 25).

The heavy ordering resulted largely because of the exaggerated concern shared by utility managers in the 1960s that customer usage might overtake their firms' capacity to meet the demand. Rather than be caught unprepared for heavier-than-expected demand, they overordered as a means of "reserving shop space."[2] This ordering behavior did not seem too unusual at first since utilities traditionally placed orders in bursts, only to realize later that they had ordered too much.[3] However, this time was different. First of all, managers had their fears of being unprepared exacerbated when Allis-Chalmers dropped out of the turbine-generator business in 1963.[4] Though the company had a relatively small share of the market, producing primarily smaller-capacity units (except for the 1,000-MW unit promised to the Consolidated Edison Company), the remaining two domestic firms still had to absorb the new business. But previous orders had already strained the resources of those manufacturers, which had begun planning construction of new plants for producing turbine-generator components. Even after the new facilities operated, however, General Electric and Westinghouse still were forced to delay shipments to utilities well beyond the two-year average that had been common between 1948 and 1962.[5]

The situation turned gloomier for utilities in the late 1960s, when the annual growth rate of electricity usage soared to an unexpected (and unplanned for) rate of almost 9% for three years. (The 9% rate implied a doubling period of only eight years.[6]) Mounting labor disputes at the manufacturers led to strikes, causing further backlogs.[7] The unanticipated growth spurts, combined with the backlogs, created panic among some utility managers, who ordered units as quickly as possible just so they might have *some* capacity when they needed it. Consequently, manufacturers found themselves literally "choked with orders" in the late 1960s[8] and could do no better than provide four-and five-year delivery schedules.[9] As *Electrical World* put it, "[u]tilities are planning, and buying, new generating capacity as though there's no tomorrow."[10] Unhappy about the delayed deliveries (as well as the prices charged for equipment), the American Electric Power Company took the unusual step of purchasing 2,200 MW of power equipment from a foreign supplier in 1967.[11]

Still worse was to come, however. When fears of capacity shortages turned into reality in several parts of the country during the late 1960s and early 1970s, requiring utilities to reduce voltage to customers in order to conserve energy (causing "brownouts"), aroused critics pointed to the need to plan for more power plants (and more reliable interconnection of large utility systems). Utilities therefore continued their ordering spree. The "run on the bank" occurred in 1973, when utilities ordered more than 100,000 MW of new capacity – over 20% of installed capacity at the time.[12] In early 1974, Westinghouse alone had a backlog of 90,000 MW.[13] One American Electric Power executive remembered that the industry felt "stampeded by the threat of not getting a piece of the manufacturing capacity. Just imagine," he added, "100,000 MW of orders and the maximum amount of manufacturing capacity for boilers . . . was 30,000 MW. There was no price competition. . . . Everybody was loaded up to here. . . . We were in a panic in the 70s. The industry got turned up on its ear."[14]

Despite being turned on its ear, the utility industry wanted more than just additional capacity. It also sought more powerful individual units to maintain its ability to produce ever-cheaper electricity.[15] As evidence of this desire, consider the ordering pattern of utility purchasing agents. In 1960, utilities placed only 10% of their orders for units in the capacity range of 600 to 799 MW and none for anything more powerful. Just five years later, however, they ordered almost 50% of the units in this range, and about 10% for units larger than 800 MW. By 1970, more than 50% of orders were made for units larger than 600 MW, with about 15% in the 800- to 999-MW range, and 20% in sizes larger than 1,000 MW.[16] In short, utilities created an agitated market for increasing amounts of huge machines.

Manufacturers felt confident they could produce the desired quantities of reliable, large-scale equipment using the design-by-extrapolation method. Unhappily for everyone, however, the vendors did not succeed. Two reasons appear to explain why. For one, the practice of engineering with lots of margin for error came under scrutiny. While it gave engineers confidence that their attempts to extrapolate would succeed (i.e., machines often could be rated at a higher capacity than for what they were designed), the extra margin cost the vendors dearly. If some of the margin could be reduced, then vendors could sell the machines less expensively (as was demanded by utilities), being priced on the basis of their actual design performance. Of course, the manufacturers did not expect that the technology would suffer from shaving some margin. After all, much of the margin had been built into the technology because analytical techniques simply did not exist for predicting the amount of stress, for example, that turbine blades could withstand. Because computational techniques for analyzing steam flow in three dimensions proved impossibly complex through the 1950s, simplifying assumptions and extra reserves had been designed into machines. But in the 1960s, as digital computers came into use, engineers believed they could tackle three-dimensional analysis and learn where margin could be shaved.

A second and more significant problem resulting from use of the design technique consisted of insufficient learning and experience. By definition, design by extrapolation discounted the value of experience. Unfortunately, it appeared to be discounted too much, especially when the period between design and construction grew longer in the 1960s. Because it took about three years to create a new design, two years to build it, and another year to install it, at least six years had passed before in-service experience could develop for a new design. In fact, experience from designs installed in the early 1960s only began accruing by the late 1960s, years after several intermediate designs had been made and machines manufactured. This "leap-frogging" approach to introducing new technology disturbed a Federal Energy Administration official, who reflected in 1979 – a time when lead times had increased to up to eight years for fossil-fuel units and ten years for nuclear plants – that there had been "little opportunity to gain operational experience with a specific design, learn its operation and maintenance drawbacks, and adjust for them in the design of a new plant." The official concluded that "plant design features may be used for several different customers before there had been any operational experience with and adjustment of that design."[17]

Here is where the design-by-extrapolation technique became apparent: even with a design strategy that seemed to work well in the past, some experience and feedback became absolutely necessary with dramatically scaled-up equipment. The use of sophisticated new tools, such as the computer, may have instilled confidence in the manufacturers. But computer programs often gave a false sense of precision because the hundreds of choices and assumptions made by programmers were not always known to the programs' users. In short, the design-by-extrapolation approach resulted in the appearance of new phenomena and problems that could not be predicted without practical experience. Pressed by their customers for more and larger equipment, however, manufacturers felt they could not allow experience to accumulate before pushing on to the next designs.[18]

Evidence of learning problems

Evidence of deteriorating quality in large-scale technology comes from a gamut of sources. Anecdotal remarks issued by utility engineers and executives offer some insight into the problems. Manufacturing engineers who worked with the technology in the 1960s and 1970s also provide a glimpse of the environment in which problems occurred with unhappy frequency. At first, people simply noted the existence of problem-laden equipment. Then came an emerging realization that the quality of power technology deteriorated when experience did not accumulate fast enough to be incorporated into new designs.

Among the earliest complaints of equipment trouble came from New York City's Consolidated Edison Company. Having had positive experience with several overbuilt and reliable older units, utility managers were annoyed after having purchased from Allis-Chalmers the largest turbine-generator set ever built in the world.[19] Just four months after being installed in 1965, the Northeast blackout shut down the entire utility system along with the pumps that circulated oil through the turbine-generator's bearings. Without auxiliary power to pump the oil, the unit's bearings burned out. Caused by the lack of backup power systems rather than by an inherent design flaw in the 1,000-MW turbine generator, this accident nevertheless marked the first of many troubles that plagued the unit. Beginning in 1968, the machine was forced out of service because of stability problems, damaged insulation, and a generator stator that needed replacement. To repair the unit, the company spent $2.5 million through the middle of 1971 alone, and this figure did not include the cost of replacement power nor the extra cost incurred when using less-efficient generating units that normally operated only when the demand for power reached peaks in the summers. But cost was not everything. The unit's outages (one lasting about eleven months) sometimes meant that the utility did not have enough capacity to supply the needs of its users, thereby causing brownouts and increased costs that distressed the utility, the state regulatory commission, and customers.[20] Such problems attended the operation of the unit that a new "joke" emerged among the utility's employees. As indestructible as the older turbine generators were, so easily destroyed was the new one: if the unit operated at 100% of nameplate capacity and someone turned on an extra 100-watt light bulb, the machine would collapse![21]

The Pennsylvania Power and Light Company suffered too. After installing several 200-MW units in the late 1950s, the company ordered machines that had double this capacity in the 1960s. Before these had been installed, however, the company ordered still more units that increased the output again. Jack Busby, president of the firm, explained in 1969 that he "hoped the new machines would run just like the old ones we're familiar with, and they sure as hell don't." Operating problems plagued the company's 900-MW units. The firm shut down one machine because of twisted turbine blades. Another operated at reduced capacity because of ash problems in the furnace.[22]

Quality problems with scaled-up equipment also spurred General Public Utilities president William G. Kuhns to complain publicly in 1969. Equipment often arrived late from manufacturers and had defects in them – problems "that simply should not have happened," he pointed out. To remedy the situation, the industry needed to "develop a level of quality control that has been applied to nuclear equipment."[23] Soon after Kuhns made these remarks, an industry journal resonated his viewpoint. "[A]ccording to buyers," the editor reported, "the quality of equipment received has nose dived to the point where both PAs [purchasing agents of utilities] and suppliers are ready to don a straight jacket."[24]

Perhaps the most damning condemnation came in 1971 from Philip Sporn, who attacked the manufacturers for the shoddy equipment that increasingly became reported by utilities. In a scathing criticism, he pointed out "that being promised a delivery and even erecting a unit and turning on steam as scheduled are not enough if the machine cannot stay on line. . . ." He further cautioned his utility colleagues that "buying and installing generating capacity and sponsoring and accepting new designs of advanced technology is something more complicated than giving a manufacturer an order and writing a contract."[25] Reinforcing Sporn's criticisms at almost the same time, a utility-sponsored task force noted that "[i]n the case of fossil-fired units, problems of decreasing availabilities, increasing forced outages, and increasing outage duration must be overcome." Work must be performed, the task force admonished, to improve the reliability of turbines and controls in steam generators, turbine designs, bearings, and other equipment.[26]

The fundamental *cause* of these problems emerged as inadequate learning by manufacturers who used the design-by-extrapolation approach. The first suggestion of this culprit came in the early 1960s when utility managers preparing the 1964 *National Power Survey* became aware of an increasing number of unreliable large units. Metallurgical problems appeared to be occurring more frequently than expected in the huge machines, and the managers wondered whether manufacturers had moved *too* rapidly to design large units. Noting that designs "were being extrapolated to larger size even before the smaller size had been proven,"[27] the managers still did not believe they had uncovered a problem that would persist. During a period of general enthusiasm for the success of their industry, the engineers felt confident that manufacturers would devote more effort to the problem and bring availabilities of new, larger plants to a higher level.

Perhaps the best indication of learning problems comes from the manufacturing managers themselves. Eugene Cattabiani, Westinghouse vice president for Power Generation from 1978 to 1986, looked back at the late 1960s as a period when manufacturers became complacent and sloppy. Their market grew rapidly, "and growth covers a lot of faults." But more importantly, the manufacturers simply tried to continue doing successfully what they had done in the past – namely designing bigger units. The technology did not necessarily become more sophisticated, related Cattabiani. It was simply bigger and produced more power from each pound of metal. Economies of scale existed, and the manufacturers tried to exploit them without concern for problems that may have arisen. According to the Westinghouse manager, they "overpushed size," exceeded the companies' "knowledge base," and scaled beyond where they had enough good test data. As a result, he noted, manufacturers "started getting a lot of problems with the turbine generators."[28]

One major problem at Westinghouse became evident as its engineers designed low-pressure turbines that had increasingly longer last-stage blades. Besides requiring extra strength for fighting increased centrifugal forces, the blades had to withstand steam passing through them at both subsonic and supersonic speeds. These problems caused tremendous analytical difficulties for designers. Engineers later learned that they had crossed an "invisible boundary" in 1967 when they jumped from using

28.5- to 31-inch blades for the last stage of low-pressure turbines. Superficial examinations of previous steps in the decade suggested that everything would work out well. But lacking full-scale – or even half-scale – test models, the engineers did not anticipate the problems they eventually encountered. Soon after installing the new turbine equipment, some customers began complaining that the units did not live up to their contracted efficiency. But Westinghouse managers did not seem overly concerned. "This should have tipped us off that we had stepped over a line that we didn't know was there," reported one Westinghouse engineer. "We thought it was a fluke – one of a kind. We didn't really take that one to heart."[29] Moreover, it was easy (and more palatable) to consider other reasons for the problems. First came the customer's inability to operate the machine properly. If used incorrectly, the machine would certainly not work as well as its design promised. And secondly, when Westinghouse engineers detected cracks in some blades, they attributed the defect to a manufacturing flaw rather than a design inadequacy, thereby masking the fundamental problem.[30]

Westinghouse managers eventually realized that the problems were self-made. In an effort to capture a larger share of the market during the 1960s, when utilities ordered huge units, they admitted to extrapolating from current designs without undertaking sufficient research efforts and developing a sufficient knowledge base.[31] The problems became so severe in the early 1970s that Westinghouse equipment was forced out of service four times more frequently than units made by GE. In retrospect, commented Cattabiani in 1986, it is hard to believe that anyone bought a Westinghouse turbine generator, noting that even if the company gave away the equipment free, the cost of repairs and maintenance combined with the expense of replacing power lost from the machines would make the choice uneconomic.[32]

At General Electric, similar problems occurred, though not to the same degree. Rapid scale-ups of generator designs in the period from 1952 to 1972 revealed major problems as units "grew" too quickly. J. T. Peters, GE's general manager of steam turbine-generator marketing, noted in 1973 that the design of scaled-up generators, for example, yielded few problems as long as the output increased in increments of 25 to 30%. "But unquestionably," he continued, the performance of generators in "increments beyond 40% was alarming." An in-house study revealed that in almost every case, equipment that had been designed to produce 40% greater output than its immediate predecessor "had major failures leading to the need for costly redesigns, costly rebuilds in the field, and the additional costs involved for purchased power."[33]

These experiences motivated GE engineers to analyze the problems' sources. Using data from 770 turbine-generator units that GE had sold (of all sizes and ages), the company found 46% of the forced outages occurred because of design problems.[34] Further analysis suggested that equipment deficiencies stemmed largely from the pitfalls of design by extrapolation, that is, from

lack of service experience . . . which cannot be avoided by imagination, mathematical analysis, laboratory tests, and logical extensions of knowledge gained from service experience. The increase in unit ratings, for example, leads to a constant need for the development of new materials, new components, and new manufacturing processes. To support these developments, extensive programs must be carried out in laboratories providing information for the design of reliable units with long lives. However, it is not economically feasible or technically possible to simulate all the conditions which will be experienced in service later.[35]

More recently, another GE manager of the turbine business group admitted that experience had not accumulated fast enough with components of larger units. Difficulties arose because of the failure to imagine what could go wrong when sizes increased. Harmonic failures in the design of turbine blades, for example, had not been anticipated and could only be remedied for later designs.[36]

In other words, design by extrapolation caused defects that showed up only after units had begun operating. These problems should have been anticipated, according to GE's Peters, because they were inevitable given the rapid leaps taken. He further noted candidly that experience is what was needed, and that "it is safe to assume that reliability and quality will increase, as the degree of extrapolations from the ratings or technology on which suppliers have direct and relevant design experience is kept under control." In addition, "reliability can be expected to increase as a function of the amount of the suppliers' available in-service experience."[37]

The Atomic Energy Commission appeared to concur when it issued a special ruling in 1973 addressing a potential problem that arose from using the design technique with nuclear plants. After watching vendors produce designs for bigger and more complicated plants in the 1960s – from 600 MW in 1965 to 1,152 MW in 1973 – the AEC pointed out that "continual increase in the size of these plants has resulted in many plant design modifications and in a large expenditure of AEC staff review effort to assure the maintenance of a consistent level of safety."[38] The rapid scale-up also aggravated the problem of learning how to build and operate a nuclear plant optimally. As a result, the agency declared a *de facto* limit to the size of new units. With power equipment then in the 1,100-MW class being installed and operated, the AEC limited future units to a thermal level of 3,800 MW. (In other words, the AEC forbade construction of nuclear furnaces that produced more than 3,800 MW of heat. Since about 30% to 35% of the heat was converted to electricity, the AEC ruling limited plants to about 1,300 MW.) The limitation had a twofold intent. First, the AEC needed to make sure it could handle administratively the increased amount of work involved in certifying the newer plants. Perhaps more importantly, however, the policy would tend to "stabilize the maximum size of nuclear power plants *until sufficient experience is gained* with design, construction, and operation of large plants."[39] To at least one TVA manager, the decree signaled the AEC's concern that the nuclear industry had moved too quickly into untested technological waters.[40]

The problems with the new technology took a toll on the manufacturers, with Westinghouse especially hard hit. Aside from turbine-blade irregularities, the company encountered difficulties with its generators. At a time when GE moved to water-cooled generators, Westinghouse stuck with gas-cooled machines, largely because the firm's engineers took pride as the first to introduce them years earlier and because they thought they understood them. But the bigger generators required more passages through which cooling gas passed, and they therefore became "spongy" in a mechanical sense. In other words, the machines lost structural strength, but several years passed before the problems showed up in the field.[41]

In an event that would have been inconceivable fifteen years earlier, several utilities sued Westinghouse for failures they experienced with turbine generators.[42] In 1974, for example, the American Electric Power Company took the manufacturer to court and asked for $75 million in damages because of a poorly operating 800-MW unit.[43] But AEP did not stand as the only problem-laden utility, requiring Westinghouse to institute an expensive "retrofit" program of fifty large turbines and eighty-five generators (in the 500-to-1,100-MW range) for other utilities at a cost of about $35 million. Partly as a result, the manufacturer's earnings in its Power Systems division declined from $86.1 million in 1972 to $61.4 million in 1973.[44] As summarized by Nicholas A. Beldecos, Westinghouse's vice president of the Large Rotating Apparatus Division in 1974, the problems may have resulted from jumps in the size of units and the glut of orders placed in the late 1960s. These simultaneous events caused "indigestion" that the company could not handle.[45]

While bearing much of the blame for problems with technology, the manufacturers should not be held totally culpable. As was admitted by TVA managers, who perhaps learned of design errors earlier than others because of their use of pioneering technology,[46] they too contributed to the decaying situation. Like other utility managers, TVA executives encouraged manufacturers to use the design-by-extrapolation approach as a way to reduce electricity production costs. "I don't think many people would quarrel with saying that we, in the industry, went up too fast in size," commented one TVA manager, noting that the utility needed more experience before moving on to next steps. "We pulled the string a little too much," he added. What the agency should have done, the manager opined, was to "get some experience and then use that experience as a base on which to move up, just like the British Central Electricity Generating Board did when it leveled off at 600 MW units and then developed experience."[47] TVA director S. David Freeman put it more succinctly in 1984: "American utilities were trying to save nickels and lost dollars," especially with their push for gains in low-cost nuclear power, resulting from their "worship of low priced electricity." TVA, he admitted, was no exception to this rule.[48]

A similar view concerning pressure put on manufacturers to extrapolate too rapidly came from a vice president at Commonwealth Edison – one of the technological leaders in the utility industry. In the 1960s, noted executive vice president Bide L. Thomas in 1985, his company shared the commonly held goal of building plants for less than $100 per kilowatt in first costs. In what seemed like an impressive feat, the firm accomplished its objective! But second costs ate up most or all of the savings that came from the cheap plants. Because manufacturers "tried to squeeze out every nickel," some equipment suffered from a lack of conservative margin or other problems that made the facilities only marginally adequate. As a result, Commonwealth Edison spent millions of dollars repairing and retrofitting several plants to make them live up to proper performance standards. "We encouraged [manufacturers] to shortchange the plants. . . . So we were our own worst enemies," remarked Thomas. "You can't just blame it on the manufacturers."[49]

In short, manufacturers contributed to the onset of stasis by using a design technique that allowed them to respond to the pressures of the utility market, but that failed to provide traditional performance and quality. To be sure, some extrapolation is always necessary if technological frontiers are to be pushed back. But extrapolations still needed to be done cautiously, with learning and feedback occurring as well. Starting in the late 1960s, however, manufacturers suffered as they tried to do too much – in terms of scale increases and margin shaving – and much too quickly. They placed their confidence in their ability to produce new technology as well as they had in the past, despite the fact that they had undertaken a different design philosophy. They also trusted their new analytical tools, such as the computer, which they felt allowed them to bypass the need for experience.[50] Thus, along with the natural barriers imposed by thermodynamics and available metallurgy, the management of a technological design strategy also imposed a limit. One engineer who admitted to having used the extrapolative method without the benefit of field experience concluded that "[w]e had designed ourselves into a box where it really looked like our technology was bankrupt. We had gone way beyond where we should have."[51]

9 Maelstroms and management malaise

Surprisingly, the limits to improvement in thermal efficiency and effective barriers to scale increase did not appear to command the attention of many utility managers. Surely, executives of companies that had purchased unreliable equipment sought remedies from manufacturers, but few seemed to recognize the implications of technological stasis for the centrally important grow-and-build strategy. After all, the availability of constantly growing and improving technology effectively drove the strategy, and stasis invalidated it. Instead, managers focused on problems endemic to the entire economy, such as high inflation, soaring interest rates, fuel-cost increases, and restrictive regulation.

The question arises: Why did these technically trained managers fail to understand how stasis affected the technological foundations of their industry? Two reasons come to mind. First, general problems with regulation and the economy posed genuinely serious crises that required vigilant attention. They certainly created management dilemmas of a magnitude not witnessed since the Great Depression – imperiling the financial stature of several utilities in the mid-1970s – and executives needed to deal with the issues promptly. The second reason is more subtle and requires an understanding of how the industry lost much of its former vigor and excitement, making it difficult to attract innovative new engineers and managers who might have been more sensitive to stasis. Those engineer-managers whom the industry could attract often lacked the imagination and critical facilities of their professional ancestors of the early twentieth century. Instead of challenging utility "dogma," they clung to assumptions and customary business practices that had worked successfully for decades, but which went *unexamined* when the industry's conditions changed in the 1970s. These managers, therefore, did not fully understand the complex intertwining of general economic malaise *plus* technological stasis. They denied that technological problems existed (in the few moments when they thought of them) and considered the maelstroms of the 1970s as aberrations in a trend toward continued productivity improvements.

Maelstroms of the 1970s

Even if power technology had maintained its traditional pace of incremental improvement through the 1960s and 1970s, utility managers still would have had their hands full. Starting in the late 1960s, the industry encountered an assortment of severe problems that tested the mettle of the best managers. Inflation posed the most obvious and immediate crisis. Resulting from increased government spending on domestic and foreign ventures (especially related to the Vietnam War), the inflation rate crept up from a sleepy annual rate of below 3% from the early 1950s through the mid-60s to an average 6.7% per year in the 1970s. At the same time that inflation hit, so did a slowdown in economic growth, causing "stagflation." After averaging more than 4% annually for three decades, growth in GNP declined to 2.6% in 1969 and averaged 2.9% per year through the 1970s.[1]

Inflation struck the industry hard from several angles. Because building new plants formed an integral part of the utility business – one utility executive called his firm a "construction company"[2] – the inflated costs of building greatly affected its financial health. The Handy-Whitman index of the cost of labor and materials going into steam power plants rose 120% between 1970 and 1979, in contrast to a rise of just 23% for the previous ten years.[3] Meanwhile, as investors sought bond yields that exceeded the inflation rate, utilities watched their old bond prices plummet and new bonds carry soaring coupon rates – reaching an average of 11.85% in December 1979, up from an already stratospheric 10.28% in September 1974.[4] For the most capital-intensive industry in the United States, the cost of borrowing became a major balance sheet concern, especially when plant construction took longer than expected. According to the lead manager of the troubled Seabrook, New Hampshire, nuclear plant, begun in 1972 and still not operating in 1989, time-related borrowing expenses amounted to 44% of the total cost of a plant built in 1973 and 87% in 1985![5] Even in the more benign late 1980s, a nuclear plant with a ten-year construction period would incur 40% more expenses than an identical plant built in just five years.[6]

Contributing to inflation, of course, was the general rise in fuel prices since the late 1960s. Coal prices surged as labor troubles forced some domestic mining companies to delay shipments and as new mining safety laws required many mines to shut down.[7] Meanwhile, oil prices moved upward as early as 1970 when the long-dormant Organization of Petroleum Exporting Countries (OPEC) exerted pressure on American companies to pay higher royalties. The 1973 oil embargo, however, made an even more dramatic impact on energy prices. Stimulated by Western nations' support of Israel in its war with Arab countries, OPEC raised oil prices from $3.00 to $11.65 per barrel in just a few months.[8] The Iranian revolution in 1979 caused a reduction of oil production and continued the upward pressure on the commodity's price, so that by the end of the decade, oil cost 441% more than it did a decade earlier.[9] Though not imported, domestic coal (especially low sulfur coal) also became more expensive because of its demand as an alternative fuel.[10]

Regulation also made life difficult for utility managers. On the federal level, passage of new laws in the 1960s and 1970s for protecting the environment put pressure on utilities to reduce pollution, often without regard to cost. A response to a heightened public sense of safeguarding the nation's ecosystem, these actions required utilities and their suppliers of fuel to do everything from reclaiming land destroyed by strip mining to controlling thermal and particulate waste pollution at power-plant sites. Environmental regulations also forced utilities to shift from dirty coal to cleaner burning oil in the early 1970s, causing some companies – especially those in the heavily oil-burning northeastern states – to watch their fuel costs skyrocket. Additionally, expenditures for environmental clean-up equipment increased as pollution-abatement requirements grew, accounting for well over 10% of the expense of the entire plant.[11] Besides its cost, the waste-management technology reduced overall productivity (defined as net total output divided by total input factors) because it consumed some of the electricity produced at the plant.[12] In the same way, environmental equipment also reduced power plants' overall thermal efficiency because electricity had to be diverted for plant use rather than for sale. In the dramatic case of the Virginia Electric and Power Company, environmental technology caused much of the decrease in the company's overall efficiency from 34.9% (9,766 Btu/kWh heat rate) in 1967 to 30.4% efficiency (11,235 Btu/kWh heat rate) in 1980.[13]

Likewise utilities received little cheer from state regulators. Once a benign and polite activity that ratified decisions already made by power companies, utility regulation became a hornet's nest as commissioners relearned the nature of their business. Often, despite understanding the financial problems suffered by utilities, regulators responded to public pressure and hesitated to pass along "rate relief" in the form of higher prices.[14] Utility companies' earnings suffered, contributing to the fall in stock prices that had peaked in 1965.[15]

As regulatory proceedings became increasingly politicized, public utility commissions caused other problems for utilities. Environmental concerns in the 1970s meant that licensing nuclear, fossil-fueled, and hydroelectric plants took longer and had more uncertain outcomes.[16] Even when approved, new plant construction times mounted due to regulatory changes made during plant construction and because of the increased complexity of the plants themselves. In just a decade between 1965 and 1975, the length of time required to build a moderately sized plant grew from about five years to more than seven years (and usually longer for nuclear plants). Because regulatory bodies in most states forbade utilities to charge customers for construction expenses (known as "construction work in progress") until after the plant became "used and useful," the delays meant mounting interest and other expenses that would create a need for drastic rate hikes when the plant eventually came on line.[17]

Finally, managers needed to contend with possible and real capacity shortages that threatened to darken several cities. Despite the rash of orders to manufacturers for new generating equipment, many utilities got caught with extremely thin reserve margins. Construction delays caused by manufacturers' late delivery of equipment, faulty installation of equipment, labor problems, regulatory clearance procedures, and environmental litigation meant that the power plants simply would not operate when needed.[18] In 1969, for example, the Federal Power Commission warned that 39 of the country's 181 major utility systems maintained less than 10% reserves for the winter ahead, prompting a *Business Week* writer to note that "utility executives have their fingers crossed that they will be able to keep the lights burning on Christmas eve."[19] The power shortage had already caused several companies to reduce voltages during peak-use summer days in an effort to lower power consumption, and future high-usage periods would also strain systems. The Consolidated Edison Company, for example, faced a severe capacity shortage in the late 1960s. Unable to gain approval of its proposed Storm King pumped storage plant and suffering from existing equipment that failed to operate properly, the company instituted voltage reductions in 1969 and continued with them for several summers thereafter.[20]

These financial, regulatory, and capacity problems occurring almost simultaneously obviously captivated utility managers of the day. Constituting immediate concerns with easily definable origins, they created real management crises. But partly because of their ease in identification, they received the most attention, while fundamental problems relating to technology remained substantially neglected.

Management decline: From high tech to low grade

Utility managers overlooked technical problems for another reason: they suffered from a mind-set that had developed over several decades within an industry that was perceived as stale and unimaginative. Managers no longer felt the excitement of the industry's early years, and they generally performed their tasks as had their immediate predecessors. Once seen as innovative, ebullient, and exciting over a period of decades, the electric utility industry gained the appearance of being lackluster – even to recently graduated engineers looking to begin their careers. Few individuals recall the large contrast, simply because they could not have been in management positions for such extended periods. But a long-term historical perspective illustrates how the industry went from "high tech" – in today's jargon – to "low grade." Unable to attract the "best and brightest" engineers and managers into their ranks, the mediocre managerial corps failed to identify and deal effectively with technological stasis.

No formal definition of "high tech" exists, but the power industry in the period before World War II shared several features with the computer, microelectronics, and other high-tech industries of the 1980s. For example, the electric power technology of the early twentieth century offered tremendous productivity improvements for those who used it properly. In factories that converted from steam-engine-driven machines to electric-motor-powered devices, productivity soared as process technology could be arranged to accommodate an orderly sequence of manufacturing events. In homes, too, the use of electricity raised people's standards of living (defined in terms of having more creature comforts). The new form of energy made it possible for people to do things that they had never done before, nor even considered before, such as listening to national broadcasts on radios and televisions. Not unlike today, when personal computers have invaded the workplace and home, electric power had a transforming effect that made people marvel.

But there were more traits that made electricity the high-tech field of the era. Besides exciting the business world and public, electricity "energized" the engineering community, creating legions of people interested in learning skills that would permit entry into the field. To meet the demand for power education, new engineering schools and university departments had to be created along with new syllabi and teaching materials.[21] Like the computer science field today, too many potential students competed for access to few formal educational channels. Meanwhile, new professional organizations sprouted to certify and provide continuing training for the new engineers. Though primarily oriented toward telegraphic engineering when founded in 1884, the American Institute of Electrical Engineers became the primary organization for power engineers by 1900. The institute's journal, *The Transactions*, provided a forum for disseminating new information outside of universities, complementing a trade journal, *Electrical*

World, whose forbears consisted of telegraph journals. Similar educational and organizational events occurred in England and Germany.[22]

Meanwhile, just as one sees in California's "Silicon Valley," Boston's "Route 128," and other high-tech areas, consulting firms emerged to cater to companies that developed and used the new technology. Stone and Webster in the United States, Oskar von Miller in Germany, and Merz and McLellan in England grew into large consulting organizations in the 1920s to assist utilities in integrating their systems during a period when regional networks became technically and economically attractive.[23] The appearance of high tech continued when the rapid growth of the industry necessitated novel types of financing to sustain it. Exploiting the relatively new idea of holding companies, utilities found a way to raise huge amounts of capital and create large systems. Similarly, high-tech companies in the 1980s use venture-capital devices to transform small ideas into big fortunes (if they are lucky). Finally, utility managers appeared daring and imaginative. Pacific Gas and Electric, for example, carried a frontier-type image in its early days, known for transporting water turbines over hazardous mountain ranges so they could provide service to a grateful public.[24] And Insull's Commonwealth Edison Company encouraged General Electric to develop novel steam turbines that eventually provided the basis for an expanding industry. In short, the electric utility industry exhibited similar types of growth, productivity improvements, manpower shortages, financial novelties, and industry and corporate daring that provided the high-tech "high" that one sees today in other industries.[25]

As the industry became "established" and mature, especially in the years after World War II, however, it forfeited the excitement that characterized it earlier and became more conservative and risk-averse. While some large companies still showed eagerness to use fancy and "neat" technologies, the perceived risks had diminished. Even Philip Sporn's "aggressive" actions could be considered relatively conservative by general business standards in unregulated industries. After all, the American Electric Power president simply pushed harder and took more risks than his colleagues for incremental advances in conventional technology that followed a pattern of improvements in the standard performance indicators (i.e., unit capacity, thermal efficiency, transmission voltages, etc.). He never took a major risk, like Insull did in 1905, by trying completely novel approaches for producing power. So, while vendors developed other technologies besides power technology (such as computers and jet engines at General Electric), utilities generally stuck with the technology that they had used for several decades.

Utility managers became conservative for several reasons. An important one has to do with the success they enjoyed with one type of technology. Why change when everything looked rosy, especially when managers had developed a satisfactory culture and value system around the mature technology? They saw prices decline, standards of living increase, and utility profits grow as they pursued a strategy relying on incremental improvements in a technology that had become familiar, understandable, and predictable. Another reason existed too. The technology required on the order of $4 to $5 of investment for every dollar of sales (per year).[26] As leaders of the most capital-intensive industry in the country, utility managers understandably resisted switching to a new technology (if an acceptable one existed) that might present too high a risk or did not offer satisfactory savings over previous investments. Thus, according to a Duke Power Company manager, doing what had been done previously was not "conservative," but "smart" absent a technological breakthrough.[27]

The industry may have appeared more daring and innovative when nuclear power appeared on the horizon. But even here, the industry exhibited great caution and initial skepticism. As General Electric, Westinghouse, and other firms began developing nuclear power for military purposes, they also saw commercial nuclear power as a sensible strategy for capitalizing on a technology developed for other customers. And yet, the utility industry hesitated at first, unwilling to become partners in an untested technology. In other words, most utility managers acted in a conservative and risk-averse fashion.[28] But the manufacturers continued performing R&D, and they offered original plants at irresistible "turn-key" prices.[29] The final pushes came when the federal government subsidized R&D and fuel costs,[30] when Congress passed legislation that limited liabilities in case of accidents (the Price-Anderson Act of 1957), and when the government permitted utilities to own nuclear fuel in 1964.[31] In an interesting case in which the vendors and government created a market where none existed before, utilities finally got on the bandwagon and embraced nuclear power as the technology of the future.[32]

Despite its adoption of nuclear power, the electric utility industry lost its ranking as "high tech." A nuclear system did not initially appear to require major new construction or operating skills, nor did its management. As noted by Philip Sporn in 1952, nuclear energy "will provide merely a new form of fuel, with the reactor taking the place of the boiler side of a thermal-electric generating station."[33] Likewise, at Commonwealth Edison, one of the earliest utilities to purchase nuclear plants, managers generally conceptualized fossil and nuclear technologies as identical, with only the steam-making equipment being different. As a result, management personnel shifted frequently from operating the two types of plants.[34] Further reflecting this view concerning the similarities between nuclear and fossil plants, some utilities spurned offers by manufacturers to build turn-key nuclear power plants. Instead, they preferred to use standard techniques – those developed for fossil plants – to save on expenses that they felt had been exaggerated in the costs of turn-key contracts. As noted in 1966 by one utility president who departed from the turn-key concept:

It is our judgment, based on some experience, that the nuclear business now has reached a point in its development where a reasonably knowledgeable utility organization can select a plant type and separate the major systems and equipment (nuclear steam supply turbine generator, condenser, reactor containment, radioactive waste disposal, switchyard) as manageable bits and pieces. That is to say, plan and build it in the familiar way most of us build our fossil-fired generating plants. . . . In short, our company now feels confident that it can apply standard procurement and construction concepts.[35]

And by arguing that nuclear power only altered the way power plants produced heat, utility managers and public relations "experts" contributed again to the image of an industry that performed its regulated duties successfully, but with little verve.[36]

The industry's conservative and dull image contributed to a novel type of manpower problem after World War II. Instead of recruiting those engineers who excelled when confronted with new problems, utilities attracted engineering graduates who felt comfortable in a stable field that demonstrated fewer technological challenges. Unlike in electric power's first few decades, when "entrepreneurs" such as Edison, Westinghouse, Insull, and Mitchell rationalized a fluid and exciting utility industry, the recruits in the post-World War II period generally did not come from the same cast. "Where are the bright engineers?" asked a 1968 *Electrical World* editorial lamenting the fact that utility companies could not attract good engineering talent – talent that would find its way into management ranks.[37] Underscoring the results of several studies throughout the 1950s and 1960s, the editorial explained that universities' electrical engineering programs before World War II invested heavily in resources for teaching about power technologies. But with the onset of World War II and the defense build up after the conflict, electrical engineering students gravitated to the more exciting electronics and aerospace industries.[38] The move left the power industry with an image of being "lethargic and plodding."[39] As a result, utilities lost their appeal as places to perform novel engineering work.

Employment statistics illustrated the trend. Whereas 26% of electrical engineers graduating from universities entered the utility business in 1946, only 11% did in 1969.[40] Even though the number of graduates rose dramatically between these years, so did the need for engineers to replace retiring executives and to deal with the needs created by rapid load growth. As noted by trade journals, however, utilities consistently failed to meet recruiting goals.[41] Faculty abandoned power engineering, too, leaving fewer universities to offer programs to students still interested in the field. In 1966, for example, the number of schools with power engineering programs in the country could be "counted on the fingers of one hand" – a sharp contrast to the more than 150 universities that offered such programs in 1935.[42] Good textbooks for use by graduate students also became scarce, resulting naturally from the lack of good professors who would write them

Figure 26. Cartoon showing engineering student discarding textbooks for 60-cycle electric power and carrying electronics books instead. Reprinted with permission from the Institute of Electrical and Electronics Engineers, Inc., as appearing in Vol. 2, *IRE Student Quarterly* (December 1955), p. 3.

and students who would study them.[43] An excellent representation of the situation consists of a 1955 cartoon portraying a slide-rule carrying, crew-cut electrical engineer walking away from a trash can that contains books on "60-cycle power"[44] (Figure 26). Suggesting the same theme twelve years later, another cartoon portrayed the utility industry as a slumbering giant. While dozing, the giant allowed the "romance industries" to steal the golden-egg-laying goose that produced all the "bright students"[45] (Figure 27).

Not surprisingly, the difficulty in attracting the best engineers to the utility industry led to the emergence of a hierarchical hiring system. According to one observer, "the electronics and aircraft people got the A students; the top electrical manufacturers got the B's; and utilities got the C's and D's."[46] As more C's and D's became entrenched in the industry, however, it became harder to attract the A students, therefore creating a self-reinforcing effect. "Electric utilities have not ranked high on the 'desirable employer' lists of graduating engineers for many years," noted *Electrical World* in 1984, with the best recruits entering the "glamor industries – electronics, aerospace, computers." It continued by noting that "this ho-hum attitude on the part of graduating engineers towards the electric utility industry has had a profoundly negative impact on the quality of its leadership – its top executives, its middle managers, and its supervisors."[47] Some managers did not worry about this trend, however, feeling that the A

Figure 27. The "Slumbering Giant." 1967 cartoon illustrating how the utility industry appeared to be losing its best potential engineers to other industries. Reprinted with permission from McGraw-Hill, publisher of *Electrical World*, as appearing in Vol. 168, *Electrical World*, August 14, 1967, p. 39.

students would not be particularly useful in the business. They might even be harmful, as was argued in 1965, because "we have little or no use for the straight A students anyway since they usually have very poor ability to get along with other folks. We pick our young fellow from about the middle of the class and give as much weight to how well they can lead others and how personable to the public they are likely to be."[48] This view appears to have predominated. In 1984, TVA director S. David Freeman still felt there were "too many C minus students in the utility industry."[49]

Besides discouraging the best engineering students, the industry's poor image also made it difficult to attract high-caliber professional managers. To many business school students, the life of a top utility executive did not seem engrossing enough, consisting of planning and financing new construction while asking utility commissions for rate decreases. As was noted by a *Business Week* journalist in 1969:

> In the past, demand was met calmly and easily. Electric utilities were quiet, unobtrusive places where corporate blood pressure was as steady as the hum of the generators. There was plenty of time for leisurely business meetings, and decisions were as frequent, simple, and neat as the clipping of bond coupons. Only the simmering feud between public power advocates and private companies gave life to a dull industry.[50]

Figure 28. Cartoon showing how engineering students were being lured away from the power industry by the promise of big money in other industries. Reprinted with permission from McGraw-Hill, publisher of *Electrical World*, as appearing in Vol. 170, *Electrical World*, October 7, 1968, p. 69.

As a consequence of this persistent view, reflected by one story told at the Harvard Business School, utility companies concentrated on recruiting the lower-ranking members of graduating classes.[51] These students constituted the equivalent of the C's and D's of the engineering graduates.

The difficulty in attracting good engineers and managers could not have been aided by the lower-than-average pay the industry offered to new recruits. A 1968 *Electrical World* cartoon tells the story effectively. As a utility manager nervously looks on, university graduates follow the steps of a pied piper who plays a tune of money and opportunity in other industries (Figure 28). Even in 1984, compensation for professional utility employees did not compare well with those in other fields, as is evidenced by the pay levels of the chief executive officers of companies. In a 1985 *Forbes* study of 785 companies in 42 industry groupings, the CEOs of electric utilities received the lowest compensations – with the median executive earning $277,000 per year. The median for computer company CEOs was $670,000, whereas aerospace and defense CEOs earned $1,000,000.[52] Combined with the industry's image as one that offered limited opportunities for advancement and little professional challenge, the low pay further discouraged the best people from entering.

Unexamined assumptions

In the 1970s, the utility industry desperately needed good engineering and management people for dealing with technological stasis and the other significant problems that it faced. But older managers – those arriving on the scene before World War II – still had high management positions, and they carried the intellectual baggage of the past era in which technological problems differed greatly. And the new managers who joined firms in the postwar era appeared less imaginative and daring, satisfied to work within an environment that promised few creative challenges. Employing previously successful business practices, they contented themselves by building new and more efficient power systems and selling more electricity. Unfortunately for these people, the use of these standard assumptions and management precepts reflected a form of reasoning termed "inductive fallacy" – the belief that the future will inevitably be like the past. With few exceptions, reliance on past beliefs and entrenched assumptions clouded utility managers' perceptions. It led them to discount technological stasis and its implications, and it exacerbated an already difficult economic and regulatory situation.

Three assumptions appear to have been especially problematic in the 1970s. The first consisted of the *assumption that technological progress would continue unabated*. Having special significance, this assumption was an element of American social culture as well as the electric power business. It derived from a popular view that mythicized nineteenth- and twentieth-century technologists and businessmen like Edison, Carnegie, and Ford as creators of the American industrial revolution. These men and others like them invented and employed novel technologies for producing goods that appeared to improve on a regular basis, contributing to the welfare of the general population. Though some people may have criticized the need for annual model changes as wasteful (especially in the automobile industry), most others looked forward to and expected this type of material progress. As noted earlier, Americans believed in continued advances in technology, which often translated into higher standards of living.

In the electric utility industry, managers had good reason to believe that technological progress would continue indefinitely. For almost a century, they had witnessed (and taken great pride in) incrementally improving technology that yielded regular productivity improvements and declining costs. Not only was it cheap, electricity also was consistently available, ready to work for the housewife and industrialist with a flick of a switch. This history of success with technology provided the basis for utility executives to assume that further advances could be expected in both fossil-fuel and nuclear power technologies.

Evidence of this belief pervades the engineering and trade literature throughout the postwar period in the form of predictions of future technological advances. Based on extrapolation from experience, these forecasts generally failed to consider events and factors that might have inhibited continuation of established trends. For example, a General Electric manager for marketing large steam turbine generators saw few limits to the improvement of equipment. Writing in 1955, he expected that initial steam temperatures would continue increasing at their historic rate of 12 °F per year, so that by 1975, new units would use steam as hot as 1,400 °F. (The actual maximum temperature reached was 1,150 °F.) He also asked and answered the question of how powerful can turbine generators get? "There probably is no truly definite answer," he noted. While economic factors and overall system design might impose some barriers, "we believe that turbine-generator units of the future will neither be limited by pressures and temperatures of the steam to the turbine nor by the generator output."[53]

Such a hopeful prediction coming from a 32-year veteran of a company that manufactured equipment might be viewed as marketing hyperbole today, but it did not conflict with other views of the future coming from less jaded observers. One engineer in 1957 looked to the industry's past and extrapolated thermal efficiencies and steam pressures for 1980. He anticipated, for example, that the average efficiency of units would increase to 36% (a heat rate of 9,432 Btu/kWh), whereas the most efficient plant would attain 45% efficiency (a heat rate of 7,598 Btu/kWh). Steam pressures, meanwhile, would rise to greater than 7,000 psi.[54] (In 1980, the average efficiency of fossil-fueled units remained just under 33%, and the most efficient reached about 40%.[55])

The enthusiasm for technological progress (and the belief that it would continue) did not wane in the 1960s and 1970s. Though noting that larger units would be more complex due to "more tubes in steam generators, casings, bearings, pumps, valves, piping, welds, relays, controls and the myriad of items which constitute a major power plant," one consulting engineer argued in 1968 that no reason existed "to anticipate that the economy of size trend will be reversed in the next 20 years."[56] He therefore predicted that the largest unit available for operation in 1976 would be in the 1,500-MW range. By 1985, 3,000-MW units should be in use.[57] Less frequently, one could find optimism that coal-burning technologies would also continue to improve. Philip Sporn, whose utility burned coal exclusively as its fossil fuel during his tenure as president, expected to improve *average* thermal efficiency of his firm's power plants from 9,457 Btu/kWh (36.1% thermal efficiency) in 1963 to 8,900 Btu/kWh (38.3% efficiency) in 1975, and to 7,500 Btu/kWh (45.5% efficiency) by 2000. While he saw challenges to meeting these goals, he did not appear to think they were unrealistic.[58] Further supporting these optimistic views, the authors of the 1970 *National Power Survey* published a graph that suggested 2,500 MW as the size for the largest fossil-fuel turbine generator in 1990[59] (Figure 29). And in 1979, Edwin Vennard, a utility manager since the 1920s, still published graphs that extrapolated units in the 3,000-MW size range for the 1990s[60] (Figure 30). (The largest unit in service in 1976 and 1985 had a

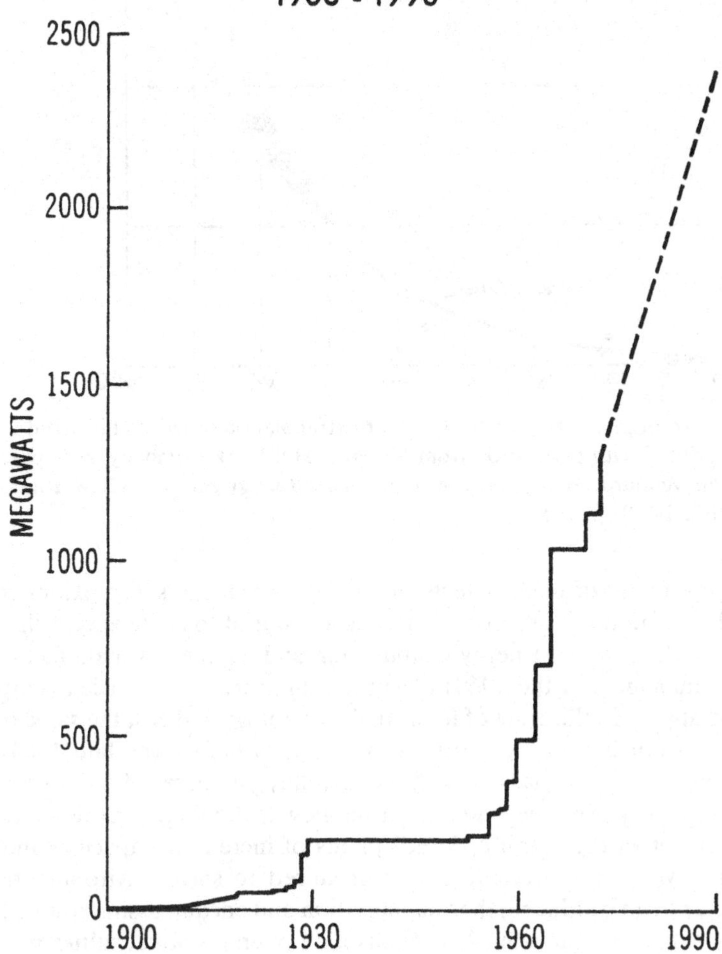

Figure 29. Largest fossil-fueled steam electric turbine generators in service, 1900–90. The graph extrapolates the size of turbine generators to about 2,500 MW by 1990. Reprinted from Federal Power Commission, *1970 National Power Survey*, Part 1 (Washington, DC: U.S. Government Printing Office, 1971), p. I-5–3.

rating of 1,300 MW.) Few people appeared to question projections such as these, and utility managers felt confident using the extrapolative approach for planning purposes – at least into the early 1970s. The inertia was simply overwhelming.

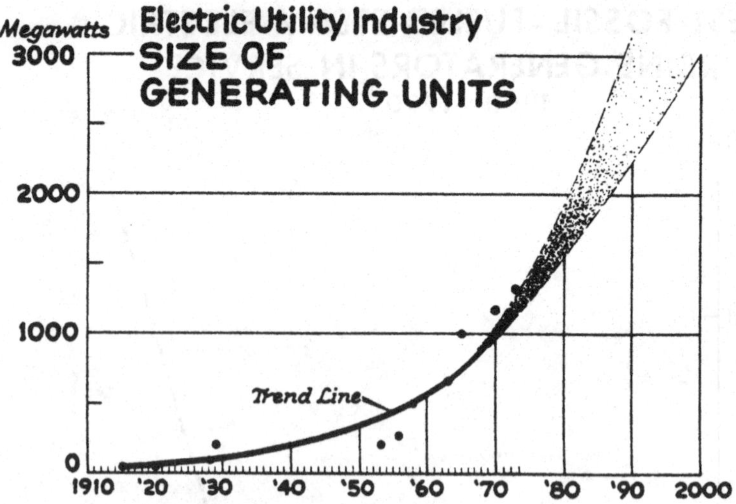

Figure 30. Size of generating units. Another extrapolation based on historical trends. Reprinted with permission from McGraw-Hill Book Company, as appearing in Edwin Vennard, *Management of the Electric Energy Business* (New York: McGraw-Hill, 1979), p. 103.

Within the realm of nuclear technology, utility managers also exhibited great faith in continued progress. Partly as a result of the extensive selling campaign of the Atomic Energy Commission and equipment manufacturers, utility managers in the 1960s began looking forward to nuclear units that would approach the scale of fossil-fuel technology and cost the same or less. Lacking much actual experience with plants before the late 1960s, utility managers nevertheless viewed the turn-key purchases of a few companies in the early 1960s as evidence of the new technology's success, and they looked forward to (and ordered) plants of increasing capacities into the 1970s.[61] Moreover, nuclear power promised to solve environmental problems. Though producing thermal pollution and encompassing risk with inherently dangerous fuels, nuclear plants did not emit smoke, sulfur dioxide, or other particulate waste. And to minimize whatever problems still existed with nuclear power, one company – Public Service Electric and Gas Company – even developed plans to build a floating atomic plant off the coast of New Jersey.[62] Thus, according to many people in government, universities, and industry, nuclear power had arrived at a propitious time in the history of electric power.[63] And despite experiences that suggested that nuclear plants cost more than originally anticipated, managers ordered them with renewed zeal in light of the unprecedented cost increases in fossil fuel following the 1973 OPEC oil embargo.[64]

The managers' view of technological progress may have been influenced by an associated belief in mechanical perfectibility. Trained mostly as engineers, managers had an educational background that stressed solving problems. This background served them well as they integrated new technologies into power production and distribution systems, but it also promulgated the opinion that "failure is but an anomaly that can be removed in a future that is more completely controlled by engineers."[65] In other words, if a problem exists, the engineer can fix it using the problem-solving approach and succeed in perpetuating progress. The view resembles the general perspective of engineers in the days of Herbert Hoover, when people believed that the technically trained man could solve technical and social problems alike. With such a mind-set, engineer-managers would naturally reject any suggestions that technology could reach limits – for theoretical or any other reasons. Consequently, when technological stasis set in during the 1960s and 1970s, managers did not generally reexamine their basic assumption about sustained technological improvement. Instead, most utility managers continued ordering large fossil and nuclear units, even though these had acquired the least in-service experience and performed the worst.

Related to managers' confidence in continued technological progress was the *assumption that manufacturers could produce high-quality technology* – an assumption that clouded managers' understanding of technical problems. Like other capital-intensive industries such as petroleum refining, the utility industry depended on manufacturers to perform the bulk of research and development that provided incremental advances and productivity improvements. Content with defining the major features of new machines, utilities generally did not do much of the engineering work necessary for building and integrating a plant full of boilers, turbines, generators, and auxiliary equipment. With final construction carried out by contracted architect-engineering firms, utilities forfeited the chance to work with the technology in which they took such immense pride. Except for large systems such as the Tennessee Valley Authority, American Electric Power, Commonwealth Edison, and Duke Power, most utilities simply had developed little in-house capability for understanding fully the detailed design of their technology. They therefore insulated themselves from the manufacturers and grew less aware and concerned about emerging technical problems suffered by their suppliers.[66] In short, utilities had become technology *users* rather than *makers* and therefore did not know about deficiencies in the manufacturers' design technique until an overwhelming amount of field experience had accumulated in the 1970s.

Utilities trusted manufacturers largely because of their long, previously successful relationship with them. Turbines, generators, and auxiliary equipment had been designed with extra margins for error in the years before 1960, and they worked well, often operating better than expected. Always eager to employ the latest and presumably best technologies for their systems, managers of the traditionally more aggressive utilities opted to buy equipment from manufacturers that *promised* economies of scale, high availability, and good performance, even though these technologies had not yet been proven by experience. The practice seemed to work well before the 1960s, and managers continued to adhere to it with later fossil-fuel and nuclear plants. Even as they entered the 1960s, a period when the utility-vendor relationship started to strain, managers still expected their suppliers to produce good equipment. At the TVA, where engineering expertise ranked among the highest of utility systems, engineers performed studies of newly offered boilers in the late 1960s, for example. While they realized that the equipment contained less margin for error, they nevertheless felt assured that the manufacturers could deliver the goods as promised.[67] No one expected that the new units would be less reliable than older machines,[68] even when the new machines had leapfrogged several intermediate designs.

As compelling as was historical experience, utility companies also trusted their manufacturers because they wanted to. After all, many utilities pursued the goal of reducing costs and customer prices (to promote growth) at a time when construction costs and inflation had begun escalating. Cost cutting had become a "crusade" at utilities such as the TVA, and managers sought all the possible means to succeed. "Often those crusades get misguided," admitted one manager in 1984, but they continued nevertheless.[69] Therefore, while their early twentieth century ancestors in utility management insisted on reliable equipment that had a foundation in experience, managers in the 1960s embraced vendors' claims and their design-by-extrapolation technique because they wanted to continue enjoying the benefits of cheaper electricity. Manufacturers insisted that they would have little problem in extrapolating to larger sizes, noted a TVA manager, and many of his colleagues gladly believed the boast.[70]

Unfortunately, as has been seen, the manufacturers disappointed utilities by selling technology that suffered due to leapfrogging and learning problems. But even as bugs in the new technology became evident, manufacturing firms generally did not stop development and learn the lessons from previous units. Instead, they plowed ahead to the next design, always hoping for performance improvements without having worked out generic problems with boilers, turbines, and generators. In a word, the previously successful vendor-user relationship had begun to break down.

The *assumption that electricity load growth brought universal benefits* required the most careful reevaluation in the 1970s. Growth had great significance in almost all industries, of course, but in the electric power industry, growth had additional meaning. Load expansion became important as a key element of the successful grow-and-build strategy that correlated increased usage with advancing technology and lower unit costs. But perhaps as important to managers, growth in usage meant a better standard of living for utility customers. This was simply part of the culture of electric utility managers.

Beginning in the late 1960s, however, a basic requirement for the strategy's success vanished as production technology stalled and as the marginal cost of electricity increased instead of declined. Facing greater expenses for construction, fuel, and other factors of production *without* technological advances to compensate for them, utility customers no longer benefited from rapid growth in usage and new construction programs. In fact, utilities began seeking higher rates from customers in order to recover higher costs. Bucking its history and culture, for example, the TVA instituted its first rate increase in August 1967[71] – two years before the average price of electricity for the nation hit bottom. Other companies followed suit. Unlike 1965, for example, which still saw eighty-three companies obtaining rate decreases of over $113 million,[72] 1969 witnessed rate hikes granted to nineteen companies of greater than $145 million.[73] And this was just the beginning. In 1972 – after fuel costs had risen about 50% since 1969 but well before the Arab oil embargo shot the price of fuels up more dramatically[74] – ninety-four companies received rate hikes of more than $827 million, while none received decreases.[75]

Even with these rate increases, utilities suffered along with their customers. Because companies still promoted growth in consumption, overall costs continued to rise sharply, especially after new, expensive power plants entered service. But regulatory bodies generally permitted rates to be based only on "historical" costs during a "test year," usually a year or more earlier for which complete data were available, rather than on the higher projected costs for an upcoming period.[76] Therefore, utilities suffered from what has been called regulatory "lag." Though they incurred higher costs, they could not charge customers for them until a later period, thereby forcing them to absorb higher costs and take a lower return on their investments.[77] Several years of rising costs and inadequate rate actions left many utilities in precarious financial positions.[78] Nevertheless, such well-entrenched strategies and assumptions as load growth could not be overthrown easily. Even as some large utilities faced brownouts in the late 1960s and as building expenses and electricity rates rose year after year, many companies promoted the use of electricity as usual. After all, wasn't increasing usage of electricity one way to lower costs? Promotion certainly was part of the grow-and-build strategy, and utility managers generally believed it should be continued.[79] In many cases, it required regulatory actions to cause utilities to *cease* promotional activities.[80]

The belief in continued growth – despite higher costs – resulted partly due to ignorance concerning the correlation of demand with electricity prices. Because electricity usage added so much value to people's lives, industry executives generally could not believe that increased prices would lead to reduced growth rates. No matter what the cost, managers thought, economic growth and public welfare depended on increased electricity usage. Especially among residential customers, load growth would not slacken, noted an *Electrical World* editorial in 1972, "because there appears to be considerable inelasticity"[81] between usage and cost. This pervasive belief set the stage for the great shock when electricity demand slipped following the oil embargo. "1973 and 1974 knocked us for a loop," remarked one TVA manager who saw residential load growth decline in the year after the embargo. The decline seemed modest – about 1/10 of 1% nationally – but it contradicted the notion of continuous load growth.[82] In 1974, utilities projected electricity usage for the next ten years would grow at an annual rate of about 7.5% – not significantly different from pre-embargo predictions. Experience proved the forecasters wrong for the next few years, however: total sales from 1973 through 1976 rose only 8.6% for the three years![83] The slow growth rate still did not move forecasters, whose expectations of future growth came down only slightly to an annual 6.3% rate.[84]

The ignorance about the effects of price on demand appears to have been heightened because the utility industry had performed few systematic studies of elasticity. Since load growth could be so easily predicted before the 1970s – one manager noted only partly in jest that "if you had a straight edge, you were a load forecaster"[85] – managers made little effort to develop computer models that considered price elasticity or other economic factors. In fact, phrases and concepts such as "price elasticity" and "long-run marginal costs" were new to the lexicons of many managers in the early 1970s, even though managers knew in general that lower prices stimulated demand.[86] Stunned by the apparent naivete of utility managers, Harvard Business School professors who convened a conference dealing with New England's power needs in early 1975 could hardly believe "the shared ignorance among some of the most informed people in the country about some very fundamental things concerning the nature of electrical energy demand." The managers appeared to know less about electricity demand, they noted, than other business executives knew "about the nature of demand for toothpaste, lifesavers or beer."[87] In short, utility executives believed that electricity usage could not be altered by increasing costs. They "saw their customers frozen into consumption patterns which they could not easily change."[88]

Another reason why managers thought growth rates would persist can be attributed to the long period of time that they had been in the business. One consulting engineer to the industry remarked that a newcomer to the industry would see the low load growth of the mid-1970s and deduce that reduced consumption had become a permanent phenomenon. The old-timers – those still in charge – looked at the events and called them short-term glitches, concluding that sluggish load growth would not continue forever.[89] At the TVA perhaps more than other places, people simply did not want to believe that the ideology of growth no longer remained valid. TVA director Freeman, who had come to the agency in 1977 after working earlier on a Ford Foundation report on energy use, noted how difficult it was to reverse attitudes in an organization that had produced low-cost electricity by promoting usage and by building more power plants. The notion that the world had changed was "not greeted with enthusiasm." Altering the attitudes of TVA workers, he continued, took many years "because low priced electricity was almost a religious belief down here, . . . believed as a matter of faith."[90] Managers from other utilities reiterated this sentiment. For example, a Commonwealth Edison manager remembered that no one thought conservation would be more than a one-time belt-tightening event, with growth continuing soon thereafter. "It took us a long, long time to get down to the growth rate [scenario] where we are now sitting, which is 2% or so," he explained. But "we were not alone in this; we all missed the boat."[91]

Managers of the manufacturing firms apparently found it equally difficult to accept radically lower growth rates. At General Electric's Corporate Research Laboratories, an economic analyst who worked for the company's Power Generation Group explained the leveling off of electricity usage in 1974 and 1975 as a combination of price hikes and the economic recession. A fundamental change in people's habits had *not* taken place, he asserted. By 1980, he predicted, load growth would be back to a near-normal rate of 6.5% per year.[92]

One can get an idea of how strongly utility managers held the growth assumption by looking at how they responded to challenges to it. One critic of the industry's emphasis on growth during a period of public opposition to environmental degradation and increasing fuel costs was Carl E. Bagge. A Chicago attorney for a railroad company from 1952 to 1964, Bagge took on a new role as a Federal Power Commission commissioner in 1964 and served through 1970. Caustically critical in what an *Electrical World* editor labeled "confrontation, Chicago style," Bagge assailed the continuing belief in the ideology of growth, in which the only goal consisted "of continually reducing costs in order to usher in the era of unlimited power – the era of the gigawatt – the electric energy economy."[93] The never-questioned belief, he argued at the 1970 American Power Conference in Chicago, had been "engendered by a monstrous sense of intellectual and technological arrogance which ignored not only the limitations of technology, but even more importantly, the limitation of the vision of its high priests."[94] The belief spurred utility managers to promote electricity usage even during a period when brownouts had begun occurring and when warning signs existed of forced load shedding and voltage reductions. "In its quest for promoting greater electric use," noted Bagge, "this industry is now obliged to expend its resources to meet a market demand which it has, in part, created while experiencing difficulties in meeting normal market demand."[95]

As can be imagined, criticism of the growth ethic did not fall on eager ears. While not disputing the existence of brownouts and the promise of more in the near future, industry spokesmen quarrelled openly at the conference about whether a crisis existed.[96] Instead, they could be heard protesting that " 'growth' need not be a bad word" and that a power crisis was "sheer nonsense."[97] These statements reflected the commonly held view that one of the most fundamental bases of the industry remained sound despite evidence that increased usage – without technological advances – annulled the benefits of the grow-and-build strategy.

In short, most people in the power industry viewed lower productivity, rising costs, higher customer rates, and reduced benefits from load growth as anomalous occurrences that had little to do with technology. They thought, perhaps not unexpectedly given their experiences, that financial or abnormal economic conditions such as inflation and high fuel costs had caused all their problems, without considering the importance of technological stasis.[98] As soon as the economy returned to normal, they believed, the industry would continue its trend toward regular productivity improvements.

By remaining attached to previous values and a business strategy that reflected those values, utility managers overlooked serious problems with their technology and focused on more manageable problems relating to economic conditions and regulation. Instead of closely monitoring their production equipment to determine whether larger scale still provided increased productivity gains with high reliability – a fundamental necessity for the grow-and-build strategy – managers allowed previously viable assumptions about progress and the ability of their manufacturers to color their assessments of technology. By failing to understand that the technological structure of their industry had been radically transformed, these utility executives also did not realize that the participants in the electric power matrix would no longer benefit from growth in electricity consumption. In other words, the managers seemed to forget that continuous technological advance was the engine that drove progress in the electric utility industry.

10
Criticisms of utility research and development

The firm beliefs and assumptions held by utility managers concerning technology underwent further testing in the 1960s and 1970s. Not only did managers make decisions about future plants based on their views of continual progress and their manufacturers' abilities during a period of economic tumult and rapid growth; they also had to defend themselves from a new set of critics who attacked the industry for its apparent inability to develop and deploy new technologies that would meet the needs of all its stakeholders. Becoming especially active after the November 1965 blackout that interrupted power to the northeastern part of the United States (and part of Canada), the criticism pointed to an industry that spent more money on advertising than on research and development (R&D).

The criticism of the industry's R&D efforts did not directly address the issue of technological stasis. Rather, it focused on the shortcomings of a system of technological development that remained largely in the hands of manufacturers. According to critics, vendors pursued their own goals and often neglected the needs of their customers. The utilities, on the other hand, performed practically no independent R&D of their own. Whether or not these attacks had validity, the dialog between the industry's critics and its leaders provides another way of examining the way utility executives thought about and managed their technology. Moreover, the utility industry's resistance to the criticism further demonstrates how tenaciously managers adhered to traditional beliefs. They felt their manufacturers could be trusted to perform the necessary R&D, and they looked to big and apparently advancing technology like nuclear power to provide the means for meeting increased demands for electricity that inevitably would come in the future.

Lack of utility R&D

During the utility industry's heyday in the early 1960s, it appeared that the manufacturers had done an exemplary job of providing the new technology needed to meet increasing demand for electricity at lower prices. And yet, not everyone was satisfied with the relationship between the utilities and vendors. One such person was Joseph C. Swidler, a lawyer whose experience in the industry dated to the foundation of the TVA in 1933. Appointed Chairman of the Federal Power Commission by President Kennedy in 1961 to resuscitate the moribund agency, Swidler supervised the writing of the *National Power Survey*, published in 1964. While noting the industry's success to date, Swidler chastised managers for their reliance on others to do essential R&D work for them. At the 1963 Edison Electric Institute (EEI) convention, for example, he criticized utilities for their meager spending on research, which he characterized as "too small and . . . too haphazard" for providing the expanding type of service the industry foresaw.[1] Compared to R&D expenditures in the chemical and petroleum industries, which amounted to 6% and 3%, respectively, the utility industry put in a poor showing with less than 1%.[2]

Swidler also inveighed against the manufacturers of power technology, whom he felt had relinquished their traditional responsibilities. Because of their interest in short-term payoffs, he argued, vendors such as General Electric and Westinghouse had not performed enough long-term "pure research."[3] While manufacturers had been conducting some work on novel generation technologies such as magnetohydrodynamics and fuel cells (two technologies that produce electricity without conventional turbine generators), "the effort is hardly commensurate with the large improvements in efficiency which some day will undoubtedly be achieved by one or more of these new methods of generation."[4] In short, the manufacturers did not perform the research needed by the utilities.[5]

Swidler's criticisms did not come at the best of times to gain the attention of utility managers. Confident after another year of pursuing better interconnection, increasing their use of bigger and more efficient technology, building new nuclear power plants, and rendering "the world's most reliable electric service," utility representatives at the convention did not urge caution and introspection. Rather, they "applauded speakers calling for an all-out, all-electric program." As reported by *Electrical World*, managers displayed a "certain sense of self-confidence, unity, and cheerfulness" and generally felt that the industry was already doing enough for itself.[6]

Nevertheless, as chairman of the Federal Power Commission, Swidler had the authority to stimulate the industry's research efforts. Soon after the 1963 conference, he called for more centralized research and for the formation of an interim organizing committee within the Federal Power Commission representing utilities, manufacturers, and universities.[7] But utility managers responded less positively than he had hoped. The poor reception arose, again, from the feeling that all was well in the utility industry and that the manufacturers had been looking out well for its interests. The responsibility for R&D rested largely on the manufacturers, utility managers noted in response to an *Electrical World* survey. Through pressure from the utilities, competition between them should be maintained to yield still better products in the future. Moreover, Swidler's suggestion for the creation of a central R&D laboratory not only appeared unnecessary, but undesirable because it might discourage manufacturers from pursuing their own research. Utilities would therefore have little to say about the content of new technologies.[8] In short, while utility managers felt that more money could be spent on R&D, they also believed that current approaches to improving technology sufficed within the current environment of continued success. After all, as one manager put it, "[p]ast methods of conducting R&D have produced the finest power system in the world."[9]

Despite his spurned initial efforts, Swidler continued to prod the industry into creating an R&D organization. Finally responding to his persistence in the spring of 1965, the Edison Electric Institute established an *ad hoc* study group that subsequently recommended formation of a research group called the "Joint Electric Power Research and Development Council." Renamed the "Electric Research Council" in April 1965,[10] the body had a mandate to identify potential technical problems and to coordinate research activities without duplicating efforts made by manufacturers. It would finance these projects with funds contributed by EEI's member utilities.[11]

The Council's creation came at a fortuitous time, just a few months before the 1965 Northeast blackout. Caused by faulty interconnection systems (rather than by a shortage of capacity to meet demand), the event opened the floodgate for new industry critics who focused on the industry's research efforts at a time when the population had become increasingly aware of the environmental implications of producing electric power. A second blackout, striking New Jersey, eastern Pennsylvania, and parts of Delaware and Maryland on June 5, 1967, did not help matters. Congress held hearings about the reliability of power systems and the need to speed cooperation and interconnection of systems for ensuring reliable service.[12] Meanwhile, the Federal Power Commission conducted its own investigation (and wrote a massive report) about the two blackouts and the dozen or more smaller cascading blackouts that occurred between 1965 and 1967.[13] As the hearings and studies progressed, newspaper and magazine editors wondered in print about the possibility of further blackouts and the vulnerability of the power network.[14] With titles like "Power Blackout – Again," "Still Another Warning Given on Power System," and "Keeping the Lights On," these editorials gave little comfort to a public that formerly took reliable electric service as a given. Having enjoyed little interference from "outsiders" after World War II, utility managers could not have been pleased with the public discussion.

Though pointing to the recent creation of the Electric Research Council as evidence that industry had begun organizing R&D efforts, utility managers failed to mollify their new critics. Aside from the Council's expenditures of a few million dollars – reaching almost $48 million in 1970[15] – the utilities still spent too little money on R&D, according to some people. As the nonprofit Council on Economic Priorities pointed out in 1972, this amount constituted only 0.23% of gross revenues and less than one-eighth of the industry's funding of promotional advertising. Quoting the President's Office of Science and Technology, the Council noted that the R&D effort accounted for a "remarkably small percentage by most industry standards," measuring less than one-tenth of the average for all American industries.[16]

Another source of criticism concerning the utility industry's lack of research efforts came from an unlikely source – retired AEP president Philip Sporn. Though cautiously supportive of carefully designed nuclear power equipment, he nevertheless wondered in 1967 whether utility managers had abrogated their responsibility of monitoring their technology and putting too much faith in their manufacturers. Providing invited comments to the Joint Committee on Atomic Energy, Sporn warned that:

> [t]here is a clear indication that the utilities need very badly to cut the umbilical cord . . . that ties them to the manufacturers, particularly the two great electrical manufacturers. To paraphrase the late Mr. Charlie Wilson [president of General Motors[17]], what is best for the electrical manufacturers, or what they think is best, may not be best for the utilities. They cannot act on faith that everything necessary will be done for them and for their best interests. There is something better than faith, and this is their own work and contributions.[18]

In other words, utilities should protect their own interests by performing an increased amount of research and development themselves.

Just a few years later, more criticism of the industry and its meager R&D efforts came from FPC commissioner Carl E. Bagge. The keynote speaker at the 1970 American Power Conference, Bagge blamed part of what he saw as an already existing power crisis on insufficient efforts by the industry to perform research that went beyond market demands. Despite past achievements, he noted, "the delegation by this industry of its basic responsibility for research to its equipment manufacturers has proven insufficient." Most importantly, he added, "the cash flow objective of your own research and the profitability criteria of your equipment manufacturer's research simply do not coincide completely with the needs of your industry and the public."[19] In short, the industry had not been looking out for itself.

The fight over an R&D organization

The greatest threat from interlopers, however, still lay in the future. As some industry critics warned of an impending "energy crisis" due to fuel shortages and construction delays that had already caused brownouts, Senator Warren G. Magnuson of Washington state introduced in 1971 an unexpected amendment to the Powerplant Siting Act. It called for creation of a new federal agency for organizing and conducting the industry's research.[20] If approved, the proposed "Federal Power Research and Development Board" would consist of five presidential appointees who held no business connection with the power industry, and it would authorize research in all aspects of producing and distributing electricity. Major goals of the Board would consist of (1) obtaining increased efficiencies in generating, transmitting, distributing, and consuming electricity; (2) decreasing the adverse environmental impact of electricity production and use; and (3) developing basic innovations for producing electricity.[21] Perhaps most unnerving to utility managers, the amendment called for the establishment of a trust fund for supporting research, with money coming from a fee levied on the consumption of electricity. For each kilowatt-hour of electricity used, a customer would pay 0.15 mills, which translated to a funding level of between $200 million to $300 million in the early 1970s.[22]

Industry reaction to the amendment came swiftly. Finally realizing the significance of the criticisms aimed at utilities and afraid of losing autonomy, the industry established its own new research arm, the Electric Power Research Institute (EPRI). Voted into existence by EEI's Electric Research Council just one week before Senate hearings began, EPRI constituted one response to a realization occurring in "about the year 1969 or 1970" that problems had arisen. According to EEI chairman Shearon Harris, managers developed the perception that "we were about to get in trouble on energy in the Nation as a whole, and that there were some real solid needs of new ways of doing things in order to keep the economy and our social order at the level at which we had arrived."[23] As described in the Senate hearings, the new EPRI would direct "research necessary in nuclear, fossil fuel, and other types of generation, transmission and distribution, utilization, systems and environmental matters, as well as to integrate and coordinate these efforts with appropriate Government agencies and manufacturers."[24] Funding for the new organization would come from the voluntary contributions of utilities at a rate that reached 0.1 mill per kilowatt-hour in 1974 and beyond, yielding about $125 million to $150 million.[25] As noted in the official press release announcing the creation of

EPRI in June 1972, this amount of funding would be four times greater than the cooperative R&D funding by the entire utility industry for that year.[26]

Critics of the industry took aim at two elements of both the proposed energy R&D board and EPRI. First of all, the utility industry could not be trusted to perform its own research and development. Not that utility managers were inherently dishonest; but due to the conservatism of utilities and the regulatory environment that sustained it, according to John Holdren, a physicist who testified on behalf of the Sierra Club, the menu of technological opportunities would remain too small. Utility managers had developed a generation philosophy, he argued, that had "built up a good deal of momentum toward larger and larger generating facilities." He felt it would be unlikely that research could proceed on small and dispersed generating sources such as solar cells in "an industry so thoroughly committed historically to the opposite philosophy."[27]

The second focus of criticism concerned the small amount of planned funding for the two research organizations. S. David Freeman, then energy policy research director of the Twentieth Century Fund, pointed out that the Electric Research Council in 1971 had called for a massive program of R&D that would total over $30 billion by the year 2000. Even with funding from utilities, government, and manufacturers, the money spent during the first few years would be $500 million per year *less* than the industry itself admittedly needed.[28]

Despite such criticisms, the Senate Committee apparently heeded the request of Charles Luce, chairman of New York City's Consolidated Edison Company. Noting that the industry had learned much about the need for more R&D efforts during the preceding few years, he urged that the committee give the EPRI a one-year grace period to initiate an R&D program on a voluntary basis. "If it doesn't develop as I think the national interest requires," he promised, "I will be one of the first to come back down here again and say the tax of the type proposed here is necessary."[29]

Luce *did* return to the Senate hearing rooms, but not to argue for the new tax. Instead, he urged continued deferral of Magnuson's amendment so that EPRI, which opened its doors in January 1973, could get underway without discouragement from a competing body.[30] Having a budget of $61 million for the first year of operations, the Institute's president, Chauncey Starr, a former dean of the Engineering School at UCLA, began establishing ties with manufacturers, utilities, and universities. Most of the research would be contracted out to industrial firms, but universities and government laboratories would also benefit from ERPI's largess.[31] Responsible to a board of directors from the utility industry and obtaining advice from a general advisory committee and from several technical committees,[32] EPRI established long-term research as its originally stated goal. Recognizing that manufacturers would not find this kind of work attractive, due to the lack of relatively near-term paybacks,[33] EPRI began to sponsor far-sighted research on breeder reactors (for commercial operation by the mid-1980s), fusion reactors (establishing their feasibility within five to eight years), and environmentally safe technologies for burning coal.[34]

The effort by EPRI's supporters appeared to pay off – at least in political terms. Congress never returned to Magnuson's amendment, presumably because the Arab oil embargo a few months later distracted attention from the power industry and diffused it onto other fragile energy systems as well. Along with measures for conservation and fuel allocation, the Congress and President debated measures that would reorganize governmental agencies and establish the Energy Research and Development Administration. Interest in creating a federal electric power R&D organization appears to have dissipated after the creation of EPRI and during the more general "energy crisis" of the mid- to late 1970s.[35]

The electric utility industry therefore escaped having a federal research and development agency imposed on it. But despite political success, EPRI of the early 1970s still embodied the traditional values and goals of the electric utility industry and pursued technological strategies accordingly. For example, the organization's leaders still believed in the inevitability and necessity of growing electricity usage, despite increasing difficulty in obtaining fuel resources. EPRI president Starr told Senators in June 1973 that an increased population in the United States by the year 2000 would require double its current needs. And because of the relative convenience and versatility of electricity, its consumption would multiply five to six times in the next twenty-seven years (yielding an annual growth rate of between 6 and 7%).[36] Assailing critics who argued for government attempts to reduce the growth rate of overall energy usage, Starr warned that these "regressive sociological steps for the future welfare of the country . . . should represent a last resort in national planning."[37]

To meet escalating electricity needs, Starr advocated the use of the same approach pursued by his colleagues, namely the development of "technological options." Reflecting the belief that "technology is a man-made resource whose abundance and variety can be continuously increased," the EPRI president felt confident that intensive R&D efforts would lead to ways to overcome the energy constraints foreseen at the time.[38] In short, ERPI leaders put faith in technological progress and the notion that creative engineering could solve problems without requiring society to make major adjustments. It was a typically American approach to problem solving, noted Joseph Swidler in earlier testimony,[39] and highly acceptable to a utility industry that looked forward to scores of breeder reactors for producing cheap, abundant, and environmentally benign electricity.[40]

To be sure, the creation of EPRI constituted an important step by the industry to assess and address important issues without relying *solely* on its manufacturers. Its founding also signaled a recognition (among some people) that the industry must deal more effectively with formerly quiescent stakeholders in the electric power matrix – politicians and the public. Despite these positive moves, many of EPRI's managers still embraced some of the same values and assumptions that contributed to the oversight of technological stasis in the first place. In other words, EPRI managers still reaffirmed the basic tenets of load growth and the need for large-scale technology to supply the increasing demand for electricity – two tenets that contributed to the industry's problems in the first place. While EPRI would eventually try to lead the industry rather than follow existing trends (in terms of developing small-scale technologies, for example), its early history suggests that industry leaders still felt comfortable with traditional assumptions and technological strategies, regardless of what "outside" critics might have argued.[41]

11 The mid-1970s: Near the bottom

"Managing an electric utility used to be a straightforward – even serene – undertaking," reported *Fortune* magazine in the spring of 1973.[1] In earlier times, supplying cheap and abundant electricity appeared to offer little challenge to utility executives, who enjoyed a charmed life. But by the late 1960s, the life of utility managers had already grown more difficult. Problems dealing with inflation, regulation, and capacity shortages had begun challenging utility managers with serious dilemmas. Combined with coal and natural gas shortages in some parts of the country, these problems gave real meaning to what some people called a "power" or "energy crisis" as early as 1970.[2] Even Congress got into the act by holding hearings and considering bills that would alleviate the crisis.[3] Abundant supplies of electricity – something the utility industry had provided with what appeared to be effortless ease for decades – looked like they were just another part of history.

Other problems loomed for utility managers as they entered the 1970s. Along with fuel costs, financing expenses for new construction skyrocketed during a period of general price inflation. No help came from regulatory bodies, which inhibited both "rate relief" and the utilities' ability to build new power plants without "outside" interference. Partly as a result of these events, electricity prices continued their upward trend, causing the chairman of the Edison Electric Institute to declare in October 1971 that " 'low-cost' power is a term that no longer applies."[4] And after the 1973 oil embargo, electricity prices increased even more dramatically, causing growth rates in usage to fall drastically from the accustomed 7% annual figure. The loss of electricity sales alarmed utility managers, who saw conservation impinging on profits. Illustrating the problem, *Electrical World* published an article in the spring of 1974 entitled "Energy Conservation Hits Utilities in the Pocketbook."[5] Meanwhile, angered over soaring electric bills, some legislators suggested (perhaps not unexpectedly) that investor-owned utility companies should be bought by state or municipal authorities, which could sell cheaper power because they did not pay dividends or business taxes.[6]

Reflecting these experiences, the investment community became leery of electric utility stocks. Prices for utility shares drifted downward from their peak in 1965, and they took a beating in 1973 and 1974. Though some of the fall coincided with the general market decline during the recession caused by the OPEC oil embargo, several electric utilities suffered disproportionately because of inflation, construction problems, inadequate rate relief, and other ills. One such firm, New York City's Consolidated Edison Company announced on April 23 1974 that it could not pay its quarterly dividend – the first time in the utility's 89-year history – because of financial and capacity problems along with reduced sales of electricity to customers.[7] Investors abandoned the stock, whose price declined 32% on the day of the announcement.[8] The event had a wider impact, however. As noted by the *Wall Street Journal*, the Con Ed action "disproves the axiom that utilities always pay their dividends,"[9] and sent shock waves through the investment community, which had previously believed in the companies' safety as investments. Consequently, investors dumped utility shares, resulting in a drop of a popular utility stock index in September 1974 to just one-third its May 1965 value and half the average value of 1970.[10] For the year as a whole, share prices of electric utility companies fell 27%. Con Ed led the pack with a 53% decline.[11] Seeking remedies for their financial problems, utility executives actively participated in Senate hearings in August 1974.[12] A few months later, a member of the Michigan Public Service Commission described the tense financial situation in a *Wall Street Journal* article aptly entitled: "Utilities Need Help – Now!"[13]

Because they were so significant, these financial and regulatory problems masked the problem of technological stasis. Thermodynamics, metallurgy, and economics conspired together to limit the efficiency by which utilities converted less fuel into more electricity. Since the early 1960s, thermal efficiency had simply leveled off. Larger power units that offered economies of scale appeared to be the logical approach to take next as a means of reducing costs, but problems emerged here too. Due to metallurgical complications, the complexity of machines, and physical limits, large units proved to be less reliable than smaller ones, thereby negating scale economies. Contributing to unreliable equipment, manufacturers exploited the design-by-extrapolation method to the extent that learning and experience could not be incorporated into new technology programs.

Even the prospects of nuclear power began to dim in the late 1960s and 1970s, as widespread criticism started to be leveled at the new technology, its users, and manufacturers. For example, people with as diverse backgrounds as regulator Joseph Swidler and former utility executive Philip Sporn attacked the industry for putting all its "eggs in one basket"[14] and for neglecting research in conventional fuel technologies, which still showed great promise.[15] Other powerful personalities, such as Federal Power Commission commissioner Carl Bagge, lashed out at utilities for their blind acceptance of nuclear power as a cure for all the industry's woes. Noting that even in 1970, not all the promises of low cost, safety, public acceptance, and technical feasibility had been fulfilled, he concluded that "nuclear power generation was not the great panacea we had envisioned."[16] These attitudes originated, noted Bagge, in a period when industry leaders felt they could do no wrong. Managers resonated this view at a 1975 conference held at the Harvard Business School, where they claimed that the oil embargo simply reinforced the need to move faster to a nuclear-electric society.[17] After listening to "the predictable litany of horrors from utility people about the difficulties of shifting from oil to coal," the Harvard professors who convened the meeting "heard the equally predictable optimism about the political and technical status of nuclear power."[18] "In the face of abundant evidence to the contrary," the professors noted, "the utility industry continues to regard nuclear power as having 'turned the corner' or having 'gotten over the biggest hurdles' in terms of public acceptability."[19] Unfortunately for the industry, however, more hurdles continued to present themselves throughout the decade. And despite the creation of the Electric Power Research Institute in 1972, no revolutionary advances that could replace either fossil or nuclear technologies would likely arrive for at least another decade or two.

The loss of steadily improving technology proved significant because it meant that, for the first time in history, the power industry had forfeited its primary means of improving productivity and mitigating difficult economic problems. In fact, as a result of stasis and the effects of regulation that required environmental add-on equipment, the utility industry's productivity fell at a greater rate than that of the overall economy. To be sure, even if technological stasis could have been averted, the industry would still have been in trouble. Incremental advances similar to those occurring in previous decades probably would not have overcome all the industry's problems. Never before, for example, had fuel prices escalated as quickly as they had in the 1970s, creating a need to raise prices with unprecedented frequency. Still, if power technology had been developed and deployed in a different manner so that stasis had not occurred, the industry could at least have looked forward to a brighter future after other problems had been resolved.

But as one business journalist noted in 1975, "it is hard to avoid a sense of despair about the utilities' future."[20] Though describing only the troublesome financial situation, she could also have been commenting on the decline in the power industry's image and management capabilities. Once viewed as vibrant and exciting in terms of new technologies and management principles, the industry had become stale and risk-averse. System building and integration had been essentially complete by the 1930s, with further work appearing to be simple extrapolation and "tweaking" to make everything work more efficiently. Even the novelty of nuclear power had been discounted to mollify once-nervous managers and the public. The industry therefore suffered when it tried to attract innovative engineers and managers who could cope with the myriad problems. And while critics chided executives and tried to focus attention on some of the industry's problems in conducting research and development on new technologies, managers continued to run their businesses as if the standard assumptions concerning technological progress, manufacturer infallibility, and the value of growth still remained valid.

Part III
Accommodating stasis

"Flexibility" – A term used by utility managers in the late 1980s reflecting the need to retain several options for generating and marketing electricity in the future.

12
Understanding values: The basis for a new consensus

This book has identified behavior of power managers as a contributing cause of technological stasis. While the hardware ran into some fundamental thermodynamic and metallurgical hindrances, it also suffered as utility companies pushed manufacturers to attempt an accelerated and unsuccessful design strategy. The implicit conclusion is that if stasis occurred partly as a result of management behavior, it should be possible to accommodate stasis through management actions as well.

Accommodating stasis must be managed on two fronts. First, managers must learn to live with stasis as it exists now. In other words, they must develop strategies that permit utilities to use current forms of technology into the near-term future (up to ten years). They must deal successfully with stasis so they can provide sufficient, reasonably priced electricity to customers and so that their companies remain financially sound. Secondly, managers must develop strategies that accommodate stasis in the longer-term future so that when they need new hardware (at the very least, when equipment wears out from old age), they can choose from a gamut of acceptable technologies.

Both forms of accommodation require managers to understand the values of the various stakeholders in the electric power matrix. As noted earlier, the participants include the manufacturers, utilities, regulators, investors, and customers. Until the 1960s, all stakeholders formed an implicit consensus concerning the technological and business system because it seemed to provide universal benefits. But that consensus began to unravel when stasis set in. As the technology no longer mitigated increases in almost all costs related to electric power production, consumers and regulators rebelled as rates escalated. Investors felt betrayed too, as they watched the utilities' financial situation decay. At almost the same time, vocal groups claiming to represent the public began viewing power technology as degrading the environment and impairing the quality of life. In other words, the commonality of interests that previously inspired the grow-and-build strategy had disintegrated in the 1970s.

Here is the crux of the matter: the utility industry suffered in the 1970s not only because managers succumbed to long-held assumptions and failed to deal appropriately with technological stasis. It also suffered because managers did not appreciate the values held by all stakeholders in the power matrix. In particular, they did not recognize that some elements of the public could acquire power to challenge utility managers and stymie their attempts to resolve problems. Moreover, managers rarely considered the public as anything more than a group of indifferent consumers who should leave the executives free to run a complex technological enterprise in whatever manner they saw fit. In other words, they did not truly appreciate the fact that their industry was a publicly regulated industry in which an active consensus was necessary, especially during unusual times like the crisis-laden 1970s.

The situation began to change by the early 1980s, however. Reeling from confrontations with their constituencies, some managers rethought their basic assumptions and learned that all participants must achieve a new consensus before a successful technological strategy can be pursued in the future.

Conflict of values: Utility managers vs. the public

Rising electricity prices in the late 1960s contributed to the erosion of the consensus between utility companies and the public. Since technological stasis precluded advances in power equipment that previously counteracted higher costs for fuel, construction, and financing, utilities needed to obtain approval for increased rates from utility commissions. As noted in Chapter 9, utilities began seeking increases in significant numbers in the late 1960s. By the time of the energy crisis in 1973, consumers had already become accustomed to regular rate hikes, though they probably could not have anticipated the increases precipitated by radically higher fuel costs. The 94 companies that obtained more than $827 million in increased rates in 1972[1] preceded 235 that obtained $3.1 billion in 1975.[2] To put these hikes in better perspective, consider the extreme case of the residential rate payer in New York City. Already paying for the costliest electricity in the country in 1969[3] – these customers became indignant when Consolidated Edison raised rates soon after asking them to reduce consumption during summer capacity shortages. By January 1971, customers paid $16.41 per month for 500 kilowatt-hours of energy – 11% more for than two years earlier. But worse was in the offing. Retaining the position as the nation's most expensive provider of electricity, Con Ed continued to receive rate hikes that pushed the cost of 500 kilowatt-hours up to $40.15 by January 1977 – 92% greater than the national average and up 171% since 1969.[4]

Public displeasure of the utilities' rate actions took many forms. After a series of power crises in the summer of 1969 added to the ire over high rates, for example, New York City's Mayor John Lindsay lashed out against Con Ed, which already had been dubbed "the company you love to hate" by *Fortune* magazine.[5] Expressing the feelings of many customers, the politician accused the utility of "gross insensitivity" and "a truly regrettable disregard for the public interest" in requesting another rate increase.[6] By the fall of 1971, customers expressed their disgust with price hikes and erroneous billing practices by writing 2,100 complaint letters each month to the state regulatory body, a volume that spurred the agency to hire extra personnel just to deal with them.[7] Outside of New York, customers also criticized utilities for higher rates. One Georgia couple went so far as to sue a utility company for falsely promoting all-electric homes as "efficient" after watching its electricity bill skyrocket.[8] Other people attacked what appeared to be unfair rate structures. At hearings before state regulatory bodies and Congress, for instance, senior citizen and consumer groups' representatives noted how most utilities continued to promote electricity usage with declining block rates. But spokesmen often pointed out that electricity had become too expensive for low-income and fixed-income residents. Using a limited amount of power, these people paid the highest per-kilowatt-hour rates and allocated a larger-than-average portion of their meager incomes for electricity bills. To remedy this problem, some advocates for low-income residents suggested the creation of "lifeline" rates – rates designed to provide electricity at low cost (even below the suppliers' cost) on the assumption that people required a minimum amount of electricity for the basic necessities of life.[9] Additionally, groups protested to utility commissions and Congressional hearings about seemingly excessive rates for residential customers (in contrast to lower rates for industrial users) and the continuation of promotional rate structures during a period of supposed energy conservation.[10] In fighting for their cause, public groups and individuals received aid from organizations such as the Environmental Action Foundation, which published booklets and other information for people who wanted to challenge utilities' rate structures.[11]

One dramatic expression of resentment flared up in New Hampshire, where the U.S. Senate Committee on Commerce held hearings in October

1976 on the high cost of electricity. With the state's Senator John Durkin presiding, the hearings focused on why New Hampshire's electricity bills had risen 83% in the previous four years.[12] Meeting with representatives of consumer and legal defense groups, the League of Women Voters, city officials, and outraged citizens, Committee members heard wide-ranging allegations of arrogance and criminal conspiracy on the part of the state's largest utility, the Public Service Company of New Hampshire.[13] The effectiveness of the state's Public Utility Commission (PUC) also came into question. As noted by Richard Mark of Common Cause, a government "watchdog" group, the PUC had instituted intricate rules that inhibited public participation and understanding of its activities. Conflict-of-interest questions arose, too, since PUC staff members could legally own stock in the utility companies on which they performed research and about which they made recommendations to commissioners.[14] In short, both the state's major utility and the regulatory body appeared undeserving of public faith. Further demonstrating the haughtiness of both groups, according to Senator Durkin, was the fact that representatives of neither the company nor the commission attended the hearings.[15]

Public antagonism over increasing rates did not occur in a vacuum. Even in the absence of higher rates, vocal segments of the public had become disenchanted with utilities for what appeared to be their intransigent position concerning environmental issues. Echoing the view that existed since the early twentieth century, many utility managers still felt that uninhibited industrial and economic growth constituted positive goals, and that increased electricity production and consumption contributed to their attainment. To be sure, power plants and factories created pollution, but for the most part, utility managers (like most other industrialists) perceived it as an acceptable by-product of modern life. In fact, most managers still held the notion that belching smokestacks (in real life and in paintings, photographs, and stock certificates) served as a proxy for progress, full employment, prosperity, and a higher standard of living for the general public.[16]

While utility managers may have retained this view into the 1960s, environmental activists became outspoken critics of it. Emerging into a powerful popular force in the 1960s, and rallied by the publication of Rachel Carson's *Silent Spring* in 1962, the modern environmental movement vocalized concerns about pesticide hazards, unhealthy air, oil tanker spills and, in general, the deleterious effects on the ecosystem of large technical enterprises. Newly created organizations, such as Resources for the Future, the Environmental Defense Fund, National Resources Defense Council, Friends of the Earth, Zero Population Growth, and scores of others, emerged in the 1960s and early 1970s and joined the ranks of already existing conservation groups such as the Sierra Club and the National Audubon Society. Collectively, they sought to inform the public, create political constituencies, and to do battle with corporations and government agencies that caused and permitted environmental degradation. A watershed for the movement occurred in 1970, when President Nixon signed the bill that created the Environmental Protection Agency. And on April 22 of that year, millions of people participated in "teach ins" and demonstrations to celebrate the first "Earth Day."[17]

This growing environmental concern was intricately related to the emergence of a more questioning attitude toward industrialists (including utility executives) who previously promoted economic expansion at all costs.[18] Unlike in the past, technology now appeared to erode the quality of life at the same time as it sought to enhance individual well-being and industrial productivity. In a 1970 *New York Times* commentary entitled "The Price of Technology," John Noble Wilford captured the spirit of Earth Day participants and other environmentalists by noting this ironic connection between modern technology and modern comforts. While citizens enjoyed labor-saving home appliances and room air conditioners, they also contributed to environmental decay because of the consequences of their selfish desires. By craving more electricity to power their machines of luxury, people had already required power companies to produce four times as much energy as they did two decades earlier. But more power meant despoiling the environment with more fossil-fuel plants "spewing fine ash and sulfur dioxide from their stacks" and more nuclear plants "raising the new problem of thermal pollution in streams and the specter of accidents releasing deadly radiation."[19] In other words, environmental degradation accompanied the quest for the modern technological lifestyle, and people would now have to make difficult choices about what they valued most.

Newly emerging public values concerning the environment found ready expression by politicians who saw a good issue around which to rally. (Even President Nixon did not want to be perceived as opposing "clean air and water.") Perhaps as importantly, however, Congressmen concerned with problems relating to the power industry also discovered a severe regulatory lacuna: except for local zoning limitations, no federal agency (including the Federal Power Commission and the Atomic Energy Commission) and few state bodies had jurisdiction over the siting and construction of power plants and transmission lines that would have an impact on the environment.[20] As early as 1968, Massachusetts Senator Edward Kennedy introduced legislation for regional planning of electric power needs that would consider environmental values. "From the Chesapeake Bay and the Long Island Sound in the East to the Columbia River in the West," noted the Senator, "citizens are alarmed at possible thermal pollution and other adverse effects from large electric power plants." And plans for transmission lines streaming ubiquitously overhead had already spurred disputes in areas of historic interest and scenic value.[21] Federal legislation appeared to be necessary, and Congress debated several bills through the 1970s.

Given their background and culture, utility managers reacted predictably to these onslaughts. One utility president, Commonwealth Edison's J. Harris Ward, resented pressures to build cooling towers at its Zion nuclear plant as a way to reduce thermal pollution. Showing little sensitivity to environmentalists' concerns, he argued in 1969 that "[r]aising the temperature of the water a few degrees is no reason for a civilization to commit suicide."[22] Other managers, meanwhile, simply discounted the significance of the environmental movement by claiming that pollution was simply an aesthetic concern. They suggested a way to defuse criticism by painting high-voltage transmission towers blue, so that they would blend into the sky, and by calling their firms "beautilities."[23] Finally, managers at the American Electric Power Company (as well as those working for other companies) argued that electricity constituted a necessary ingredient for *cleaning up* the environment. While the generation of electricity created some pollution, noted the 1971 AEP *Annual Report*, "electricity in its utilization undoubtedly makes a greater contribution to the creation and beneficent environment for mankind than any other single factor" and "is absolutely essential in the fight against pollution and to the health and welfare of all people."[24]

Perhaps more distasteful to utility managers than the general attacks from environmentalists was the argument for the need to reduce the growth rate of electricity usage. If the rate declined, as some environmentalists advocated, the need for new plants (and the subsequent siting and pollution problems caused by them) would diminish.[25] But to utility people who had been brought up with the view that growth only had positive connotations, the suggestion could not be accepted. As has been noted earlier, few power executives before the 1973 oil embargo felt that growth should not remain at its historic 7% per-year rate. When Philip Sporn raised questions in 1970 concerning the immense growth in future power needs, for example, he received predictable criticism from utility representatives and the Edison Electric Institute, which noted that increased electricity usage helps keep America's industries "modern and efficient." Moreover, if the country hoped to solve social problems such as poverty, job opportunities must be expanded, with electricity playing a crucial role.[26]

The clash between the public and utility officials did not remain an academic debate. Public representatives advocating new rate structures and concerns for the environment took advantage of administrative law rules to win sympathetic hearings from state regulatory commissioners. Like utility customers, regulators had been viewed as silent partners in the implicit consensus, supporting the grow-and-build strategy as long as marginal prices declined. And while critics argued that regulators often become "captured" by the industries they regulated, commissions began upsetting utility managers in the late 1960s as they discovered a new environment existed for everyone. The old assumptions of low interest rates, cheap fuel, rapid economic growth, and improving technology had disappeared, while new environmental issues (and energy conservation concerns after 1973) had arisen. As utility managers tried to grapple with these problems, commissioners attempted to perform their dual responsibilities of protecting the public and ensuring the financial integrity of utility companies.[27] Often, however, utilities got short shrift.

State regulatory action took several forms. In many cases, for example, public service commissions simply disapproved the full extent of rate hikes requested by power companies, disappointing utility managers who felt that the commissions yielded to political forces and represented the public's immediate interests only.[28] Perhaps more disturbing to managers, a few utility commissions tried to impose innovative approaches to encourage conservation – the opposite of what utilities had been doing for so long – especially after the energy crisis had put greater pressure on rate increases. Some regulators even insisted that utilities reevaluate their grow-and-build strategy and give up the declining block rate structures that previously contributed to its success. In Wisconsin and New York, for example, state commissions instituted so-called "time-of-day" rate structures for some customer classes. Based on marginal cost principles, this mechanism charged different rates for using electricity at different times during the day or season: when the utility system's load approached peak capacity, customers would be charged more than during lax times (such as on weekends or late at night), when the company had spare capacity. The approach aimed at increasing utilities' load factors and, simultaneously, reducing the need to build new and expensive power plants to meet growing peak demand.[29] Though watered down from an original version as part of President Carter's national energy program, the Public Utility Regulatory Policies Act of 1978 carried this move away from promotional rate structures one step further: it eliminated them altogether as one way to encourage conservation of energy resources. The law required state regulatory bodies to work with utilities and other interested parties in developing new rate structures to meet this goal.[30]

Regulatory approval of power plant sites became another point of contention.[31] Before 1969, when no state had any law requiring site approval for proposed plants, utility managers decided on new locations for plants based primarily on technical and economic concerns. But new statutes that gave numerous state agencies or public service commissions the right to review plant siting made the task more difficult for managers. With twenty-seven states having such laws by 1976,[32] managers now needed to be concerned with environmental factors as well. More importantly, they had to acquire permits and licenses from a series of bodies, such as state departments that oversaw environmental protection, natural resources, fish and wildlife, forestry, water control, and so on.[33] At each step, interested parties that disapproved of a plant had the opportunity to challenge a site permit, thereby delaying or cancelling a utility's plans.[34] To some people in the industry, a fundamental inequity existed in this system. Whereas utilities had to win every battle with state agencies, their opponents only needed to win one in order to postpone or defeat construction of a plant.[35]

Finally, utility managers found they no longer could receive assured acceptance of their construction programs nor a guarantied return on their investment, even when previously approved by the assortment of regulatory authorities. Managers at Consolidated Edison experienced this earlier than most. Deciding in 1955 to build a nuclear power plant at Indian Point (on the Hudson River, 40 miles north of New York City), the utility watched the costs of the 275-MW unit increase from the $55 million estimate to $127 million when completed in 1962. At the time, conventional plants cost about $190 per kilowatt of capacity, but the Indian Point plant cost more than $450 per kilowatt.[36] Partly because of the excessive cost, the New York Public Service Commission in 1967 withheld more than $100 million from the company's rate base. According to a nuclear physicist who testified against Con Ed in a 1967 commission hearing, the company had simply been "imprudent" in its development of a pioneering technology.[37]

This type of prudence review by regulatory bodies, called to judge whether utility managements acted properly in deciding to build new power plants in the past, reached a crescendo in the 1980s. In the striking case of the Long Island Lighting Company, for example, the PSC held in 1985 that $1.4 billion of the $4.6 billion spent so far on its nuclear power plant at Shoreham, New York, could not be included in the rate base.[38] The unrecovered money constituted expenses that should not have been incurred if prudent managers had made proper decisions, the Commission ruled. Therefore, instead of utility customers bearing the cost for management incompetence, the stockholders would be responsible.[39] The action contributed to the poor financial health of the utility, which almost went bankrupt in 1988. In an arrangement made with New York State, the utility vowed to close the plant (which was opposed for safety reasons by the governor and others) in return for rate increases that would keep the company financially secure.[40]

As can be imagined, most utility managers did not approve of what appeared to be a massive intrusion of outsiders into their business through the regulatory process.[41] While many publicly expressed the need to use electricity more efficiently and to strike a balance between environmental protection and economic growth, some managers did not act as diplomatically. Trying to reassert their dominance, a few executives argued that the "uninformed" and untrained public simply could not understand the complexities of their business. One such utility executive, Theodore J. Nagel, a senior vice president of the American Electric Power Service Corporation, wrote in 1978 that complex power issues required specialized training and that decision making could not "carried out in an open forum or in the atmosphere of a town hall." He advocated the imposition of new rules for limiting the duration and scope of public hearings so that irrelevant issues would not come up. "The alternative," he concluded, "can be nothing less than confusion and chaos."[42]

Gordon Hurlbert, president of Westinghouse's Power Systems Company, seconded this argument concerning the public's lack of expertise and right to participate in making decisions. An engineer (though not a utility manager) and graduate of the Harvard Business School, Hurlbert used stronger words than Nagel in 1979 and urged that utility managers fight back against the "uneducated and uninformed dolt who can . . . [receive] a better hearing than any utility executive." He advocated a heavy hitting campaign proclaiming "Nuclear Power or No Power" as a way to counteract efforts by "all the stop-progress deep thinkers" who tried to halt projects such as the breeder reactor. The power industry should become more activist, he lashed out in an article addressed to utility managers, because

> We are men who *know* that there is a direct relationship between available, affordable energy and the economic health of the nation – and the world. We *know* that kilowatts can be converted to bread and heat and infinite forms of useful work. But we have sat quietly while that same energy has been decried as a polluter, a luxury, and a rip-off.[43]

While many readers of Hurlbert's call to arms may have harmonized with his view, the proposed public relations offensive never came to fruition. Thirteen days after the article's publication, the accident at the Three Mile Island nuclear plant occurred – an event that further eroded the once-powerful stature of utility managers.

Aside from the general mayhem caused by public intervention in regulatory proceedings, many managers specifically disliked the imposition of new rate structures and conservation measures. Feeling that once the energy-crisis and inflation "glitches" were overcome, growth and new building would again be desirable. Managers therefore stood adamant in the early 1970s against changing traditional rate structures. Reflecting their ire, *Electrical World* described as a "murder trial" a tortuous 15-month-long hearing in Wisconsin concerning efforts to modify the standard declining block rate structure.[44] And while they talked about conservation, many managers apparently disliked the notion of encouraging customers to use *less* electricity. Quoting a utility analyst in 1980, a *Wall Street Journal* article reported that managers "gave [conservation] lip service, but putting dollars into conservation goes against their grain; they still want to play with their erector sets."[45]

Finally, managers objected strenuously to the right of commissioners to convene prudence hearings. In the managers' view, regulators used 20–20 hindsight to penalize actions that had been taken in good faith and based on the best information then available. Moreover, many decisions made by them had already been approved by the same regulatory bodies that now criticized them. It seemed unfair, thought the managers, that their companies should now be penalized for plans made at a time of 7% annual growth rates in electricity usage and when most other utilities in the country had made similar projections.[46]

In general, managers resented the public's intrusion into the decision-making arena. More than anything else, they felt the public did not appreciate their years of efforts to bring down the cost of electricity and provide a commodity that offered social and economic benefits to everyone. Now that unusual economic circumstances had arisen, they thought the public had converted their apathy (something managers could live with) into scorn (something they felt was unjustified). Moreover, public groups had become serious forces with which to contend. Entering into regulatory proceedings and pressing for new laws that affected utility activities, the public (and the regulators) appeared to have become a new set of managers who could dictate terms and conditions for power companies in a way that had never been possible before. In short, managers felt they had lost their autonomy – a disheartening discovery to many. Accordingly, managers thought that the social contract – one that had served everyone well for decades (in the managers' view) – had been dissolved.[47]

But by trying to disregard or "write off" the public's initiatives in the 1970s, many utility managers made a costly mistake. Though referring only to the conflict between the public and utilities concerning erection of hydroelectric power plants, one historian nevertheless characterized the situation well when he noted that "[u]nhappy clashes with aroused groups of ecologists have proved that when a [hydroelectric] dam is being proposed, kingfishers may have as much political clout as kilowatts."[48] Instead of trying to rebuild a consensus, managers continued to believe in the ideology of growth and maintained their conviction that only they *knew* what served the public the best. As engineers, most power managers felt that several of the problems facing the industry could be solved by using *more* technology, not less. When faced with electricity shortages in the early 1970s, for example, utility managers often suggested building more power plants rather than encouraging behavioral changes among customers, such as conservation practices (through new rate structures or marketing approaches).[49] And, of course, many of those new plants would be nuclear power plants, which would live up to earlier promises as soon as the public understood them better.[50]

As demonstrated by industry critics such as FPC commissioner Carl Bagge in 1970, however, the public simply refused to "understand" nuclear power. Nor did it complacently accept the position of authority that utility managers had assumed over the years. Fighting back in the regulatory hearing rooms and in Congress, the public converted its new sense of values into law and regulatory precedent. And increased public pressure on federal agencies and state regulatory bodies made it much more difficult for utilities to construct and operate nuclear power plants. The mishandling of the Three Mile Island accident in 1979 – just in terms of providing accurate and consistent information to the public – did little to encourage better feelings.[51] Despite the hopes and wishes of utility managers, therefore, the general public radically changed its values and views toward the utility industry during the 1970s. Things that once meant "progress" and an improved standard of living – as did promotion of electricity usage and big technology – now meant "menace."

Shifting values of utility managers

The confrontational attitude between utility managers and their constituencies began to diminish in the 1980s as managers of a few leading utilities altered their approaches toward producing and selling electricity. But the "transformation" of utility managers did not come easily nor in a uniform fashion. The three examples that follow offer case studies of how different utilities dealt with novel situations that caused them to rethink their strategies.

For the managers of the Pacific Gas and Electric Company, the "conversion" experience was nothing less than traumatic. An industry giant, with about 10,000 MW of capacity in 1970, PG&E entered the decade with old strategies intact. Expecting load growth to continue at its traditional rate of 6 to 7% annually,[52] the company foresaw the need to construct five new nuclear power plants for use in the 1980s at a cost of about $13 billion. But unlike in the past, when the company usually had little trouble getting approval for its plants from the state's public service commission, the utility faced the ardent opposition of an "intervener" – the Environmental Defense Fund (EDF). This environmental organization at first appeared to pose little threat. Like other similar conservation bodies, this one would surely argue on the basis of a "smaller-is-better" philosophy that did not address directly the immediate electricity needs of Californians.

The EDF surprised the company, however. Avoiding the rhetoric of the conservation movement and using the language and accounting tools of the industry, the organization argued that the company could better serve the interests of customers and itself by encouraging conservation, alternative energy suppliers (such as windmill farms), and by using load-management strategies. This last suggestion appeared especially promising. Referring to the manipulation of customer demand by economic or technical means,[53] load management would permit the utility to meet demand for more electricity without building as many new facilities as in the past. It worked by offering incentives (such as time-of-day rates) to customers who spread out their electricity usage throughout the day and who avoided times of peak demand.[54] Though never winning in the hearing rooms, the EDF's five-year confrontation (often with the news media providing sympathetic coverage) eventually convinced the PG&E that the approach demanded serious consideration. To be sure, slower growth rates after the 1973 energy crisis helped motivate the company into accepting the EDF proposals. But most importantly, the utility's managers felt compelled to abandon many of the cherished beliefs of the industry's culture – mainly the growth ideology – and it tried to make ends meet with what it had available.[55] In the late 1980s, the company was known as among the most innovative for its efforts in conservation and load management.[56]

Not all utilities had to be confronted by aroused public and environmental groups. For example, managers of the New England Electric System

(NEES), a holding company of several utilities in the New England states, did not follow this adversarial approach. Instead, it developed a new strategy in the late 1970s after failing to obtain a license for locating a proposed nuclear plant. The disappointment meant that the company's managers needed to develop an alternate plan to meet projected electricity demand. After much internal debate, the company eventually developed its "NEESplan" in 1979 for load management and conversion from oil to coal as an energy source. Aiming for load growth of only 2.4% per year, the firm contracted for power from small, independent generators of electricity, such as industrial companies that simultaneously produced steam and electricity in a process known as "cogeneration." It also purchased power from firms that burned garbage and produced electricity as a by-product. Even with load growth of about 5.5% per year in 1983 and 1984, the company reaffirmed its goal of keeping load growth down to 2% for the subsequent ten to fifteen years.[57] NEESplan II, unveiled in 1985, envisioned more conservation, load management, and cogeneration along with purchases of excess power produced in Canada. The company also employed rate structures that conveyed proper price signals about conservation to customers, and it planned to extend the life of power plants already in use.[58] While executives in other utility firms are skeptical that NEES can succeed with its plan in light of economic expansion and higher-than-expected growth rates, the approach nevertheless signified a major change for a leading utility company.[59]

Perhaps the TVA best represents a utility system that shifted its management values to harmonize with values held by customers and others.[60] Having developed a culture that deified the notion of growth, the Authority continued its strategy of building new plants into the 1970s and 1980s. With a capacity of about 20,000 MW in 1972, the TVA planned to add another 20,000 MW by 1983 – all of it nuclear after completion of one last coal-burning plant in 1973.[61] But as energy prices escalated in the 1970s, economic activity in the region stabilized, and the need for the additional capacity dissipated. Most importantly, new directors replaced those who had supervised the authority's growth period, and they realized that conditions had changed. Cost hikes and less need for new capacity resulted in two actions: cancellation of nuclear plants already under construction or in planning stages, and a massive conservation program.

According to one observer of TVA policies, the major force in changing from growth values to conservation values was S. David Freeman, who became a TVA board member in 1977.[62] As already noted, Freeman had been a critic of nuclear power and an avid advocate of conservation. Appointed to the TVA post by President Carter, he sought to reverse some of the Authority's policies. Soon after his arrival, the TVA instituted programs for home insulation and industrial energy conservation. It also offered low-cost loans for these efforts and encouraged using alternative energy sources. By reducing the need for building new generating facilities, the TVA hoped to keep the cost of electricity as low as possible.[63]

The dramatic reversal of the TVA from being the "yardstick" of promotion to an example of conservation suggests that its managers' values changed dramatically. It was illustrated again by the admission of Hugh G. Parris, TVA's power manager, who pointed out in his 1982 annual report that by "the early 1970s, the utility industry had essentially reached a plateau in major technological development. . . . It was not possible to 'produce' our way out of the problem as we had in the past."[64] For a manager of a company that had developed a culture around the grow-and-build strategy to admit that the technology had leveled off and would not mitigate economic problems clearly illustrates the realization that the world had changed. Equally as telling of changing values was director Freeman's assessment of the nuclear power "debate" between the public and utility companies. Looking back from the vantage point of 1984, he noted that

> [t]he whole [utility] industry had blind faith that nuclear was another form of power. They [utility managers] had grossly underestimated the difficulties of the nuclear technology and the safety problems. It was a massive blunder of the technological elitists. There was a worship of nuclear scientists. It's a situation where I think contemporary history shows that the common sense of the ordinary citizen who was objecting to nuclear power . . . has turned out to be closer to the truth than the collective judgment of GE and Westinghouse and the Joint Committee on Atomic Energy. . . . Participatory democracy came up with a better result than the feudalistic centralized decision making process.[65]

The cases of PG&E, NEES, and the TVA are not offered as examples of how utilities should accommodate stasis. After all, these firms altered the standard grow-and-build strategy only when faced with major public resistance, failure to obtain approval for a building program, and replacement of "old-line" managers with new executives. The changes caused serious dislocations within the companies, and many utility managers still feel that the new approach will not serve the best interests of power companies long into the future.

Nevertheless, the use of conservation and load-management techniques have, at least for a period of time, reflected better the values of the public and the regulatory bodies through which the public expressed itself. Building programs have been severely curtailed, reducing one upward pressure on electric rates and relieving concerns over the environmental consequences of selecting new sites for power plants. As noted by American Electric Power chairman W. S. "Pete" White at his company's 1981 annual meeting, "it would be the best of all worlds if we didn't have to build."[66] The actions also appear to have appeased another group of participants – investors – as utility companies have begun to adopt the new approach of load management. Because utilities such as PG&E and NEES, along with others like Wisconsin Electric Power and Pacific Power and Light, require less new equipment and construction to meet slowly growing demand, they avoid the financial burdens associated with building large generating stations.[67] Some of these companies became "cash cows" – simply producing electricity and taking in revenues without major construction plans – and they can distribute extra earnings as dividends to investors or move into diversified and unregulated business.[68]

To many of the participants in the power matrix, then, the conflict of values between the public and some utility managers has been resolved in a way that has yielded a new consensus. Rather than being concerned only with providing a growing supply of electricity, several utility companies have used load-management techniques to live with the technology at hand instead of pursuing new and bigger hardware. Providing several years of relative stability, the strategy has allowed many companies to refrain from constructing new plants and to offer prices to consumers that rise no faster than the inflation rate.[69] Not all companies (or their customers and investors) are so lucky, of course. A few remain saddled with the results of decisions made decades ago (such as those utilities that are still trying to gain regulatory approval for nuclear plants). Still, on the whole, the stakeholders in the power matrix can celebrate a redemption from the oppression they suffered in the mid-1970s.

13 The search for new technology

The participants should not celebrate too wildly, however. Even if the new consensus based on low growth and load management becomes widely accepted, it will not solve the utility industry's problems altogether. After all, slow as it might appear compared to earlier periods, load growth in electricity usage continues at a 2.5% annual rate in the late 1980s, and new power technology will ultimately be needed. And when units wear out from old age, they will require replacement – no matter how low the growth rate becomes. Consequently, new technology must be developed for the industry, and some of it may be able to penetrate barriers that previously appeared to exist.

From a "theoretical" point of view, overcoming stasis is possible, since stasis has been defined as resulting from a combination of technical, economic, social, and managerial events. This definition is what makes the notion of stasis more appealing than that of technological maturity and other concepts that dovetail with the notion of technological life cycles. (See Appendix A.) In all these biological models, it is generally assumed that once a technology reaches the stage of plateau or maturity, no more progress ensues, with senility or death following shortly. This book discounts that view.

History provides a good example of how the utility industry overcame a form of stasis that resulted from general economic and managerial causes. During the first thirty years of the twentieth century, progress in increasing the scale of turbine generators followed a roughly exponential course, with single-shaft units growing as large as 160 MW by 1930. The next twenty years, however, saw a flattening of the trend, which could have been interpreted at the time as a technological plateau, even though thermal efficiencies continued to improve. In fact, however, market, social, and political forces (i.e., the Great Depression and the diversion of resources toward marine boilers and turbines during World War II) acted to produce this period of stability. As has already been discussed, the new market and technical conditions after the war enabled the resumption of progress without revolutionary advances. In other words, technological stagnation resulted from several causes not related to inherent technical conditions.

Ironically, even if stasis can be overcome in large-scale conventional technology, the new hardware may not be considered appropriate by all stakeholders in the power matrix. Huge and reliable power units, once deemed highly desirable, might not be well accepted by investors, leaders of utilities, regulators, or the public due to financial and regulatory risks. Accommodating stasis might therefore require development of new technologies that do not directly address the problems of limits to thermal efficiency and economies of scale. Rather, overcoming stasis might be accomplished by sidestepping those technologies that evidenced barriers in the first place.

Improving conventional technology

As contributors to the onset of stasis, manufacturers of power technology play a major role in overcoming it. Fortunately, they have worked hard to remedy technical problems during the difficult years beginning in the mid-1970s and continuing into the 1980s. The key events that stimulated a search for solutions consisted of: (1) a horrendous economic, social, and regulatory environment that slowed the pace of electricity sales, providing a break in ordering new large-scale power units; and (2) a sense of crisis that made manufacturing managers devise new strategies to ensure high quality and reliability in their technology.

The financial and regulatory problems of the 1970s, as already noted, stifled the utility industry and provided a "breathing spell" in which its managers were forced to reassess their programs of technological deployment. For some companies late in the decade, conservation and load-management programs replaced electricity promotion strategies, and ordering for big, new plants – and certainly nuclear plants[1] – diminished significantly. In 1979, for example, orders for new capacity declined to about 5% of the peak in 1973 (Figure 25). And because load growth appeared in the 1980s as if it would not increase as rapidly as in the past, the pressing need for large-capacity additions did not seem as imminent.

Not all participants welcomed this pause in ordering big plants. The manufacturers, for example, feared the event and saw much of their traditional business evaporate. To Westinghouse managers, especially, the slow period of sales provoked a crisis situation. As the number two producer of turbines and generators, the company reached the late 1970s with few orders on the books and a bleak future. One seriously considered option consisted of leaving the turbine-generator business altogether, since new sales of any importance appeared far in the future. Yet under new management, the Steam Turbine-Generator Division took a different tack in the late 1970s. Instituting a massive cultural change (which included shifting the site of its turbine-generator headquarters to Florida) and employing "Japanese-style" attitudes about quality assurance, Westinghouse began a program of designing new components of large power plants and installing them in already existing units that needed to be repaired or upgraded. The approach contrasted the older, classical strategy of designing all components for entire turbine generators, an approach that no longer proved viable when few utilities ordered complete units. But it provided a way for the manufacturer to gain experience with the new parts so that when (or if) utilities ordered new plants later in the 1980s or 1990s, the firm would be in a position to offer more efficient, more reliable, and better tested technology.

Essentially, Westinghouse used the poor period of sales to refine design techniques and to improve technology already at hand. In the process, the company overcame technical barriers that contributed to stasis, such as the problems experienced with long turbine blades. As an example of this new approach, Westinghouse was able to improve dramatically the performance and useful life of some early-1970s-vintage turbines by replacing critical parts with newly designed, "ruggedized" components. The new parts enabled the utility to increase the units' power output by 3.7 MW (for each 684-MW unit), lower the heat rate by 50 Btu/kWh, and extend the longevity of the low-pressure turbines an extra thirty years. Requiring only six to eight weeks downtime for the utility, the retrofit work offered an ultimate "payback" of over $17 million at a fraction of this cost.[2] At the same time, Westinghouse also gained experience that it could employ in designing still newer components. In short, the new approach really was not so new after all. It consisted simply of a return to the previously successful design-by-experience strategy.

The strategy appears to have overcome one major obstacle that contributed to technological stasis – lack of reliability in components for large turbine generators. As is evidenced by forced-outage statistics, Westinghouse had real trouble with the equipment it shipped between 1965 to 1969, well before the company used the new approach. Experiencing a forced-outage rate of 9% after the first year of service, Westinghouse equipment caused lengthy downtime for utility repairs and expensive purchases of electricity from other sources. In the 1975-to-1980 period, however, Westinghouse's new approach helped reduce that rate to only 2%. And for the few machines sold in the 1980 to 1984 period, the rate further declined to about 0.5%.[3] Though the new design-by-experience strategy does not guarantee Westinghouse success in the future – the press reported in the mid-1980s the possible demise of the Steam-Turbine Division[4] – it is nevertheless a novel and instructive one for American firms dealing with an old and no-longer glamorous technology. It reflects a daring attempt to overcome stasis in a period of poor sales – a period that would have been considered with horror if forecast a decade earlier.

General Electric appears to have pursued a similar strategy for improving old equipment using new techniques – and gaining experience at the same time. Never having the same degree of trouble with equipment as Westinghouse, GE still developed programs (such as its Facilities Life Extension plan) to refurbish components in power plants that had already put in long years of service. Though financially depreciated, many parts in these plants could be replaced or retrofitted so that their performance approached those of the best modern units. Benefits accrue to the utilities because they postpone investment in a totally new plant, and they can avoid the tempestuous procedures of locating and gaining approval of new plant sites.[5]

As a result of work performed during the slow years of the 1970s and 1980s, then, the major manufacturers have essentially overcome some symptoms of stasis. Though hindered still by practical limits to thermal efficiency, they have developed means to ensure better performance in very powerful units. Using new analytical techniques, R&D managers and engineers surmounted problems that previously caused reliability to suffer. More importantly, perhaps, they garnered experience with new components in retrofitted equipment. In fact, experience with these designs has been so gratifying that some managers talk confidently about building 1,500-MW units – a size that would surpass any that existed in the late 1980s. No utility may order such a big unit in the immediate future, but managers say that when the demand for large-scale technology returns in the United States or abroad, they will be prepared.

Creating unconventional technology

The fact that manufacturers of conventional technology have been working to overcome stasis is good news. But problems may still exist in the future because conventional technology of the same type and scale may not be appropriate for building a new consensus among participants. Even if manufacturers could build reliable and efficient power units that produce 1,500 MW each, would any utility want to purchase one? The answer in the late 1980s is "no" due to the hostile reception utilities would receive from customers, regulators, and investors. Meanwhile, because of the vagaries of customer demand – in 1986, an *Electrical World* editorialist called load forecasting "a pyschoanalytical crap-shoot"[6] – big units might sit idle or underutilized, creating further financial burdens for companies and investors. Regulatory commissions might even consider the underused plant to have been the product of poor management and refuse to include all its cost in the rate base. In short, big plants in the late 1980s incurred too many risks.[7]

As a result of these concerns, utilities have increasingly tried to develop strategies that encourage "flexibility." Though the term has become a new buzzword in the industry, it nevertheless reflects the view that most managers do not want to commit their companies to construction programs that take decades to complete and that might be opposed by their stakeholders. They need, instead, technologies around which a consensus can be built. In the late 1980s, this means that utilities want to get by with as little new construction as possible so as to avoid the need to raise capital and seek regulatory approvals. Utilities also desire power units that can be installed quickly and in small increments in order to ensure their quick inclusion in the rate base. For these reasons, the majority of fossil plants planned for installation before 1992 have been in the 300–700-MW range – well below what once had been considered the optimum size.[8] Econometric studies suggest that this actual practice makes good sense. After balancing many factors, the "ideal" size for a new conventional (base-load) unit has been calculated to be around 400 to 500 MW.[9] A similar view has emerged for nuclear plants, too.[10]

But conventional technology, albeit smaller in size, is not the only technology being developed. The traditional manufacturers, and perhaps more importantly, new players in the field, have been working on different technologies that may better fit current needs. Among the new players is the Electric Power Research Institute. As noted earlier, EPRI had been created in 1972 under pressure from the federal government, and it claimed to serve as a "catalyst and manager of research" on problems that are of current interest to utilities – EPRI's "clients."[11] Giving up its original long-term technological goal of developing large nuclear breeder power plants, EPRI has sponsored various R&D programs that could lead to smaller and quickly installed power sources.

An example of this type of technology that EPRI has supported for replacing conventional technology is the fuel cell. Developed in the laboratory for use in spacecraft since the 1950s (before EPRI's existence), the fuel cell converts hydrogen-bearing gases (such as natural gas or synthetic gas made from coal) into electricity without combustion. With EPRI's aid, Westinghouse Electric Company and United Technologies Corporation in the United States, along with several firms in Japan (independent of EPRI),[12] pushed the technology into the demonstration stage by the mid-1980s. Units of 4.5 MW have been installed in New York City and Tokyo with modest success. Commercial production of modular units in the 10-to-

20-MW size range may begin in the mid-1990s, with a major advantage consisting of their ability to be installed quickly and in small enough units so that tremendous capital outlays over long periods can be avoided.[13] In other words, when a utility needs more capacity, it could simply install a few more 10-MW modules in a short period of time. As was recognized as early as 1960, the modular feature of fuel cells "would represent a virtual revolution in the thinking and practices of executives of principal utility companies."[14]

EPRI has also been active (along with several companies working independent of the Institute) in developing the integrated gasification combined-cycle (IGCC) plant, another new, modular technology that shows great promise. Actually, the technology includes elements that are not new at all. The system takes advantage of gas-combustion turbines (similar in theory to large aircraft jet engines) that already exist or more efficient ones that are near commercial availability.[15] Connected to generators, the turbines produce electricity in the standard fashion using natural gas as the basic fuel. But as demand for electricity increases, the utility would add a "bottoming-cycle" system for capturing waste heat from the gas turbine to be used for powering a small steam turbine and generator. Finally, when (or if) natural gas becomes difficult or too expensive to acquire, the utility would add a coal-gasification unit – a system that takes coal and converts it into a rich gas for use by the gas turbine.[16]

The IGCC system offers several advantages to utilities. First of all, part of the technology (the gas turbine) is already familiar to utility managers and engineers, thus making it more attractive than untested alternative technologies. Next, it offers the "magical" quality of flexibility and modularity: for meeting short-term needs quickly, a utility would install a gas turbine first, which could be built from prefabricated units in as little as one year. Later, the utility would add the bottoming-cycle turbine for further capacity. Meanwhile, it would take advantage of a fuel that in the late 1980s is cheap, but that can also employ coal (in a clean version) when gas is no longer available. From other points of view, too, IGCC looks interesting. It can be installed in increments of about 100 MW to as large as 300 to 500 MW and at an estimated capital cost (in 1985 dollars) of about $1,500 per kw, approximately the same as a conventional fossil-fuel plant.[17] IGCC technology appears "on the doorstep," with several pilot and demonstration plants operating successfully in 1987,[18] and with actual orders for units coming from two utilities in the same year.[19] Meanwhile, manufacturers of the equipment, such as Westinghouse, have been working on these systems since the 1970s and are promoting them heavily. In the eleven combustion turbine-cogeneration systems (just the turbines and heat-recovery elements) already installed by 1987, overall thermal efficiencies reached 50% or better. The company expects this percentage to increase to about 75% in the near future.[20]

Beyond EPRI, several small and large manufacturers have worked to develop and promote cogeneration technology, which is being reintroduced after having been fairly common early in this century. Popular in Scandinavia and other parts of Europe, a cogeneration system consists of a relatively small steam or gas turbine-generator system (usually smaller than 150-MW capacity), and it appeals to industrial enterprises that require heat for processes such as papermaking or petroleum refining along with large amounts of electricity. The company produces electricity as would a utility, but instead of discarding the waste steam to a river or the air, it uses the heat in manufacturing or processing. It therefore reaps an energy bonanza because it uses approximately 80% of the energy content in the fuel; electric power systems convert only about 35% of the fuel energy.[21] As an example of the benefits proffered by cogeneration, consider a plant built for Pfizer, Inc., in Southport, North Carolina. Engineered, financed, and constructed by Cogentrix, Inc., a firm created in 1983 that developed standard cogeneration designs with the assistance of the Duke Power Company, the facility produces 110 MW of electricity and 945,000 pounds of steam per hour. Except for 5 MW of electricity used to operate the plant, Cogentrix sells all the coal-fueled electricity to Carolina Power and Light Company and provides the steam for Pfizer's production of citric acid. Perhaps most noteworthy, the plant came on line in August 1987, just fourteen months after construction began and at a cost of $84 million – or about $800 per kilowatt of electric capacity.[22]

Beyond fuel cells, integrated gasified combined-cycle units, and cogeneration systems, of course, are technologies that have been touted for a long time and that are still being developed by large industrial firms and small, entrepreneurial companies. Often viewed as near technical and commercial feasibility are such technologies as wind turbines, geothermal generators, and solar electricity panels. All these promise the feature of relatively small size and the ability to be installed quickly as peak demand slowly creeps up – thus providing the flexibility that utilities desire. While these have been successful in limited laboratory and pilot systems, development continues with their advocates' hopes that they will become more economically competitive with existing technologies.[23]

Can stasis be accommodated by sidestepping it with new, small-scale technologies? Such a question is difficult to answer because it depends on the continued support of the equipment manufacturing companies, EPRI, utilities, and outside research organizations. As for the traditional manufacturers – GE and Westinghouse, for example – the outlook is optimistic for gas turbines and cogeneration equipment because the firms hope to sell the technology as replacements for the large-scale fossil-fuel and nuclear plants they formerly marketed. The move seems appropriate given the weak market for conventional power plants: in 1987 (as in 1982) not a single order came in for a complete turbine-generator unit.[24] Moreover, much of the alternative technology requires little more than a modification of already existing equipment and manufacturing facilities.

The outlook for success with new technologies through the efforts of EPRI appears mixed. While the research institute has made noticeable progress with integrated gasification combined-cycle technology, which appears on the near-term horizon, it has been less fortunate with the fuel-cell project, a longer-term R&D venture. Despite years of funding and the existence of small demonstration units, fuel cells still have an uncertain future. According to the EPRI manager who administered the program, the technology has been "just around the corner" for so many years, and it is still "too expensive to get . . . to the right point on a learning curve" to be economical for utilities. The manager felt that manufacturers simply "burned out" trying to bring the technology to a commercial level.[25] This view is further reflected by the departure of General Electric from research in the field of advanced fuel cells in 1984. As another EPRI manager noted: "there's an awkward cost transition between R&D and commercialization. An automated assembly plant would bring the unit cost down, but the manufacturer needs orders to justify that investment. So far, high cost is keeping the order book empty."[26]

Another potential problem at EPRI is its overall inability to pursue long-term R&D projects on novel technologies. EPRI's budget is not very large – remaining at its 1984 level of $325 million (current dollars) in 1987 – nor is it devoted to long-term research any more. Having shifted soon after its creation, EPRI budgeted 61% of its contract expenditures in 1987 for near-term research.[27] EPRI's first president, Chauncey Starr, attributed the change in research emphasis to several factors: "the oil crisis, which quickly raised fuel costs; a need for energy conservation and increased end-use efficiency; and the increasing cost of new capital, which made it economic to extend the life of older plants and equipment."[28] But Joseph Swidler, one of the major forces behind the Institute's creation and a member of EPRI's advisory council from 1973 to 1980, attributed the change to other factors. Because EPRI depended on contributions from utilities to conduct work, it responded to the perceived needs of its utility constituency.[29] Most utilities showed interest not in long-term efforts to build fuel cells, but in minimizing pollution from burning coal.[30] Thus, some of the alternative technology programs may not be "pushed" as much as will be necessary to make technologies such as fuel cells a commercial reality.

Outside research organizations also play a major role in the success of these new technologies. While independent companies and research institutes contribute funds for R&D on new technologies (the Gas Research Institute, for example, works with EPRI as a major supporter of the fuel cell), a large share of resources comes from the federal government. During the Reagan administration, such support had not been as forthcoming as previously. Aside from the elimination of several tax incentives for solar, wind, and other alternative technologies, the government scaled back funding for R&D efforts. During fiscal year 1980 (during the Carter administration), the Department of Energy budget for developing nonnuclear electric technologies peaked at $718.1 million, up from $436.2 million in fiscal year 1977. By fiscal year 1983, however, the budget had dropped to $244.3 million. In fiscal year 1986, the agency requested only $153.9 million for nonnuclear R&D.[31] This reduced level of spending for R&D makes near-term commercialization of the new technologies less likely.

Ironically, even successful development programs may be resisted because the technologies could disrupt the established utility industry. For example, EPRI has supported research on small, modular technologies such as fuel cells that could be rapidly built and introduced into the rate base – a major attraction when long leadtimes can cripple utilities financially. However, in developing such technologies, the organization might help destabilize the utility industry by removing one major economic rationale for regional monopoly control of electric power, namely, the ability of the utility to exploit scale economies from large technologies and produce power less expensively than potential competitors.[32] If EPRI's work on small-scale systems succeeds, then it will have provided convenient technologies that nonutility users could use to produce electricity for their own employment (in a manufacturing plant, hospital, or apartment complex) and reduce their dependence on centralized utilities. As a possible consequence, the utilities that support EPRI's work on small power technologies might withdraw funding. In short, the research organization helps create technological *opportunities*, but not all will be embraced within the *traditional* structure of the utility industry.

Rerationalizing the utility industry

Perhaps, then, the utility industry simply needs to be restructured. After all, if utilities can no longer convince their stakeholders to support construction of large-scale technologies, then a restructuring may be required to accommodate the use of unconventional or small-scale technologies. In other words, the situation in the late 1980s may require the industry to be rerationalized so that the slow growth in demand can be met in a way that all stakeholders can support.

In actuality, a restructuring of the industry is already underway – and for many reasons that stem from earlier discussions in this book. Most importantly, the utility industry appears to have lost the technological imperative for remaining a monopoly. Because low growth rates combined with financial troubles and public resentment against "big" technologies militate against building large plants, utilities prefer to add capacity in small, quickly installed modular units. But these technologies do not require the giant consolidated utility companies to finance and build them. In fact, cogeneration plants, for example, can be easily constructed and profitably operated by large industrial firms, as we have seen. Moreover, it appears that small-scale producers can supply electricity at the same or lower cost as can utilities. These facts suggest that monopoly control of power generation may no longer have the same rationale as when utilities employed the largest-scale technology possible to provide universal benefits.

This emerging notion received an important boost with passage of the Public Utility Regulatory Policies Act (PURPA) of 1978. One of five laws that emerged from President Carter's National Energy Plan, the initiative had the goal of conserving energy by electric utilities, establishing fair rates to consumers, and optimizing the efficiency of producing power with available resources.[33] To achieve the last goal, the law encouraged companies to take advantage of cogeneration technologies that produced both steam and electricity for industrial uses. It also offered incentives for small power producers that took advantage of renewable resources such as solar energy, wind, water, or waste materials. Most importantly, if a company produced more electricity than it needed from any of these methods, PURPA required that a regulated utility purchase it at a rate equivalent to its own cost of producing incremental power.[34] But the law encompassed even more. Because it exempted qualified cogeneration and small power producers from most state and federal regulations,[35] it created a class of producers that existed *outside* the realm of normal utility operations.[36]

The effect of PURPA on industry structure has already been significant. Though perhaps an unintended consequence of an effort to improve the efficiency of the nation's energy use, the law has effectively begun to restructure the utility industry by deregulating it. Suddenly, the industry includes a new group of electricity producers, exempt from normal regulation, that has an almost guaranteed market for its product. The new producers also have the privilege of entering markets at any time, and they can withdraw from them as easily because they have no legal responsibility to serve (as do regulated utilities). These flexible provisions have attracted numerous industrial cogenerators as well as other energy "entrepreneurs" who see opportunities for economic gain in this endeavor. By early 1987, for example, this new PURPA-inspired class of producers accounted for about 24,000 MW of capacity, or 4% of the nation's total.[37] As they gain more experience with cogeneration technology (and as companies such as General Electric, Westinghouse, and Cogentrix promote it heavily), independent producers will become bigger participants in the power business. In 1988, for example, the North American Electric Reliability Council projected that nonutility generating companies would account for 22% of the planned construction during the following decade.[38]

The emergence of some form of deregulation in the utility industry has spurred academics to examine further restructuring scenarios. Much of the work concerns the creation of a totally deregulated power-generation industry and the establishment of independent (and somewhat regulated) transmission and distribution systems. To some observers, this means that existing utility companies must be divested, with new firms created that simply generate electricity (from coal, nuclear, other fuel resources, or from a mixture of them). In some scenarios, the new companies (and other companies that had their start with PURPA) would sell to an instantaneous spot-market system in which buyers of electricity (the new transmission and distribution firms) could choose from a gamut of suppliers (the generating companies) in order to obtain the least expensive power. (Such an idea is already the subject of an experiment among several companies supervised by the Federal Energy Regulatory Commission.[39]) In theory, when the utility industry feels the competitive pressures of the marketplace, economic and energy efficiencies will result.[40]

To many utility managers, the prospect (or existence) of a mixed utility industry made up of regulated and unregulated operators has been greeted with horror. In fact, some utility managers foresaw the threat of the new generating companies and tried unsuccessfully to overturn the law that permitted their existence.[41] (Challenges to the law received hearings by the Supreme Court in 1982 and 1983.[42]) The resistance reflects some very reasonable concerns. For example, if too many industrial firms produce power for themselves, utilities would be left serving the remaining residential and commercial customers – those that traditionally have lower load factors than industrial customers and that are more expensive to serve. In other words, the utilities would be losing some of the diversity that previously held down overall costs. Utility managers also feel that unregulated PURPA producers could decide at any time to withdraw from the electricity-producing business, thus leaving utilities "holding the bag"[43] and jeopardizing system reliability. Finally, PURPA requires utilities to purchase electricity from independent producers even when regulated firms have sufficient capacity to meet demand or when they can provide electricity at lower costs. In both cases, customers could suffer by paying more for electricity than if the utility provided it alone.[44]

These valid concerns notwithstanding, perhaps the greatest reason for resistance stems from utility managers' desires to retain control of an industry that they feel has served its stakeholders well. To be sure, the industry has faced some problems in recent years, but over the long run, haven't managers consistently provided a necessary commodity at reasonable prices? Most utility executives would answer "yes," and they generally do not want the industry's structure to be altered. Rather, they would prefer to see the industry retrogress – returning to the "good old days" when financing and construction of large base-load (preferably nuclear) power plants could be accomplished easily without outside interference. As was pointed out by an *Electrical World* editorialist in 1987: "Cogenerators and qualifying producers may prove to be dependable sources of capacity over the long run. . . . But compared to the certainty of adequacy and long-range reliability of supply offered by base-load nuclear and fossil units – well, you can use it, but. . . ."[45]

Unfortunately for many managers who think like this, public pressure and new laws such as PURPA make a return to 1965 impossible, with the result that utility managers are indeed losing control of their industry.[46] Meanwhile, a long history of relative stability and sameness has blinded many managers to the fact that in a changing world of technology and business, the structure of industries sometimes does change. This is exactly what is already occurring in the electric utility industry, and like a similar change that occurred in the deregulated long-distance telecommunications industry, managers must realize that they need to change as well.

Though few in numbers, some utility managers have begun to realize the benefits of a modest degree of restructuring. Executives of the New England Electric System, for example, recognized that the decentralization of electricity generation can proffer benefits. Like managers of Pacific Gas and Electric and other firms that have made long-term contracts with unregulated generating companies, they realize that a new industry structure can help them avoid costly construction of new power plants. By 1982, NEES had entered into contracts for purchase of 63 MW of outside power and had plans for 200 MW by 1995. Also in 1982, PG&E had contracted for 1,289 MW of capacity, of which 133 MW had already been delivered.[47] By early 1987, contracts for 9,000 MW had been written, and 1,600 MW had been constructed.[48]

Perhaps the boldest exponent of restructuring *within* the industry is William W. Berry, board chairman and chief executive officer of Dominion Resources, the holding company parent of the Virginia Power Company (formerly Virginia Electric and Power Company). Since 1981, Berry has irritated colleagues by advocating deregulation of power generation as a means to provide electricity in the most efficient manner.[49] Opposing PURPA's avoided cost-pricing mechanism that he feels leads to inefficient and wasteful transactions, Berry argues that utilities should retain the job of forecasting electricity demand and new capacity needs. Regulators would oversee these activities to ensure that companies did not develop self-serving plans. Utilities would then invite bids from cogenerating firms or other electricity producers (excluding affiliates of the local utilities) and evaluate them solely on the basis of obtaining the best deals for customers.[50] Besides price, utilities would evaluate factors such as generating reliability, ability to dispatch units in times of greatest demand, and fuel diversity. Under PURPA's avoided cost-pricing rules, Berry argues, these considerations do not receive proper attention.[51] If permitted to incorporate these factors into their planning, utilities could benefit by purchasing power from new sources without suffering the system-reliability penalties that could arise under PURPA's rules.

To be sure, Berry feels that in such a deregulated scenario, his company would prosper. For several years, his company has striven to reduce its costs of production so that it could compete with prospective newcomers to the industry. In its diversified, unregulated businesses, the company has even invested in thirteen plants to produce 2,524 MW of power outside its service area.[52] But perhaps more importantly, Berry's advocacy of deregulation indicates that at least one manager has recognized that his industry has changed. Technological stagnation, economic and financial hardships, and a restive public have all contributed to the end of business as usual. If utility companies are to survive at all and serve the public well, Berry might argue, then they will need to alter their strategies and industry structure accordingly. Perhaps he also feels, unlike many of his colleagues, that change is inevitable and that utilities' interests would be better served if they have a hand in the way the new industry is rerationalized.

Part IV
Conclusion

14
History and the management of technology

Concluding his 1959 essay on "Growth and Development in the Electric Power Industry," Philip Sporn optimistically expected historical trends in technological development to continue:

> Although the industry in the past quarter century has made very substantial technological strides which have given the country perhaps the finest series of systems for making available to its economy an abundant and highly economical supply of electric energy, many technological challenges loom up for the quarter century ahead. . . . Much remains to be done to improve even further the efficiency of generation, transmission, and distribution and to extend the field of application of electric energy. The history of how this has been accomplished, and of how challenges have been met and responsibilities discharged by the industry and by its technicians, engineers, and technologists should make most exciting and stimulating reading in 1984.[1]

This book has been written in part to demonstrate that the reading in 1984 was not as exciting and stimulating as Sporn had hoped. Generation efficiency did not increase much on average, and while transmission and distribution efficiencies did improve, they could not mitigate the high costs associated with new construction, fuel, and environmental protection equipment. Instead of reaching new peaks of productivity and performance, the electric utility industry in 1984 had fallen from its exalted position among customers, investors, and regulators.

A history of success . . . and unanticipated events

Certainly, Sporn was not an unperceptive man, and this quotation of his hopeful comments in 1959 is not meant to belittle his substantial technological and managerial skills. However, he failed to foresee why power technology would reach barriers to improvement and why this event would contribute to the industry's decline. His failure can be explained by considering several factors. First of all, Sporn wrote at a time when the utility industry still had intact a comprehensive system of technology, business principles, and culture. The basic technology of the industry had been developed by the manufacturers since the 1910s. With the help of "progressive" companies such as Sporn's American Electric Power Company, incremental technological improvements occurred regularly. Throughout the century, thermal efficiency rose gradually, providing utilities with a way to continue the trend toward lower cost electricity. At the same time, the scale of power plants exploded – especially in the post-World War II period – giving utilities the means to reduce the cost of capital invested on a unit (per-kilowatt) basis. Largely because of these gains, the electric utility industry constituted a true declining marginal cost industry that demonstrated tremendous productivity enhancements throughout the twentieth century.

Technological progress provided the basis for utility managers to develop important business principles that guided expansion of the industry. The grow-and-build strategy, for example, depended on new technology that evidenced gains in thermal efficiency and scale economies. As part of the same strategy, managers employed promotional rate structures, such as declining block rates for residential customers, to help propel the lofty 7% annual growth rate for much of the century. At the same time, utility managers developed a culture that portrayed themselves positively as people who understood and loved their technology and who provided a versatile and useful commodity to a grateful public. As usage of electricity grew from year to year, managers felt pride that they had contributed to their customers' rising standard of living and their companies' economic health.

The beauty of the technology, management principles, and culture was that they all reinforced each other and led to "progress" that all participants agreed upon. During much of the century that Sporn witnessed directly as a power manager, the stakeholders in the electric power matrix had formed an implicit consensus about the technological system and its management. Benefits accrued to all: consumers enjoyed electricity whose unit price declined gradually. Investors profited from steadily increasing dividends and share prices of utility stocks. Managers congratulated themselves for their aptitude in running a complex technological enterprise and for improving the financial picture of their companies. Manufacturers happily took new orders for the advanced technology that they pioneered. And regulators sat quietly on the sidelines, providing little interference in what appeared to be one of the best examples of a natural monopoly. It was an elegant system, and Sporn cannot be criticized for believing that it would remain intact.

But in his optimism for the future, Sporn failed to anticipate that the technology would reach apparent limits to improvement. Supercritical boilers could provide thermal efficiencies of up to 40%, and the AEP president (like many of his colleagues) naturally believed that further improvements could be made. In the early 1960s, however, he became aware of metallurgical problems that might hinder attainment of higher steam temperatures and pressures than those that existed in 1959. Nevertheless, he hoped that manufacturers would overcome this problem through renewed R&D efforts. Sporn eventually became disenchanted with the manufacturers, especially when metallurgical problems showed up in components of the huge units installed in the 1960s and 1970s. But in 1959, he did not expect that utilities (his own included) would pressure manufacturers into developing a new and ultimately unsuccessful design strategy – the design-by-extrapolation technique – that encouraged scaling up of power components before much (if any) experience had accumulated. Using the approach, manufacturers produced unreliable equipment that reversed the trend toward scale economies in large units. Altogether, limited thermal efficiencies and scale economies meant an end to productivity improvements and declining marginal costs. From the late 1960s on, utilities found that they could no longer mitigate the myriad of financial and regulatory problems with improved technology, and they watched their costs and prices increase.

These last consequences of stasis contributed to the process of destroying the implicit consensus that had lasted for so long, something else that Sporn did not foresee. Consumers and regulators became irritated when prices started rising and when a new "consciousness" concerning the environment led to recriminations against utilities. Investors, meanwhile, sought better opportunities when utilities' financial instruments lost their traditional safety and conservative status. Throughout all this flux, utility managers apparently did not fully understand what had been occurring. For the first time in history, tested business practices concerning technology, their manufacturers' ability to produce improving equipment, and growth no longer could save the day. Unwilling to discard these business precepts immediately and thinking hopefully that the problems they witnessed would be short lived, managers became even more aggravated as they lost autonomy to other stakeholders in the power matrix. Neither Sporn nor anyone else predicted profound events such as these in 1959. And who could blame them given how powerfully these experiences conflicted with those that occurred during the previous seventy-five years of utility history?

The need to understand one's history

Can utility managers learn anything substantive from this history? Moreover, are there lessons that managers of other technological enterprises can learn so they can develop a sensitivity to stasis and anticipate it in other situations? The answer to these questions, I believe, is "yes," with the primary requirement being that managers develop a better understanding of their industry's history. Because I am an historian, this nostrum should not seem so unusual. But my advice for comprehending an industry's history goes beyond simply recognizing and avoiding past mistakes. I also argue that historical events provide the framework in which management decisions about policy are made.

Though perhaps an unusual argument to be seen in the business literature, the assertion that history affects policy should not seem so remarkable. After all, in the political policy-making realm, it is accepted that historical experience constitutes an important factor for consideration. Few people would deny, for example, that the experiences of two world wars have influenced decisions arrived at in the Soviet Union and the United States in the same way that the Holocaust experience continues to help fashion foreign and domestic policy for Israel. In short, historical experience should be considered significant to policy makers.[2]

In the case of the electric utility industry, historical experience proved important for what it accumulated in a specific and concrete way: a matrix of entrenched assumptions and values upon which business decisions were often made. For utility managers, three assumptions in particular became especially troublesome in the 1970s when they lost their validity. The first consisted of a belief in technological progress that would continue indefinitely. Because progress had always been assumed in the utility industry, few people questioned the existence of technological limits. If they used any formal method of technological forecasting on which to make these evaluations, they employed what can be called the "trend extrapolation" or "curve-fitting" approach. Dependent on historical experiences, the approach had the advantage of avoiding "the uncertainty of making detailed predictions about the future of specific devices."[3] It constituted a simple way to project into the future – and a way that seemed to work in the past, not only in the realm of technological forecasts, but also in predicting growth in electricity sales.

But the consequences of retaining this belief can be severe. As noted in this book, making business policy using this assumption at a time when technology is no longer improving simply leads to financial and regulatory trouble. Moreover, when managers believe that technology will continue to improve in small increments, they become insensitive to any of a myriad of circumstances that could cause stasis or simply make the next steps much more difficult. In the power industry, this implicit belief had become widespread, with managers sharing the popular notion that since small technological advances worked fine in the past, so similar moves in the future should have similar outcomes. But the risks of incremental steps were not always the same, especially when engineers tried to make progress when performance indicators appeared to be flattening. For example, the risk of moving from about 38% thermal efficiency with conventional boiler-turbine-generator systems to 40% with supercritical units were not the same as previous 2% increments. Problems arose as the novel equipment failed to operate as reliably as standard equipment or as well as manufacturers had hoped.

Consider another example of the failure in this type of reasoning: in the late 1950s and early 1960s, many utility managers believed that nuclear power constituted an incremental advance in power technology. It constituted simply another, more cost-effective way to boil water. Aside from the boiler, the rest of the power technology would remain basically the same. Of course, the reasoning proved fallacious as nuclear technology demanded different engineering and management skills to meet safety and reliability criteria. The lesson is that (contrary to conventional wisdom) history may *not* repeat itself, and that promised incremental advances should be considered carefully to determine whether they will incur new or greater risks.

Even if limits to technological development do not appear imminent, reliance on historical patterns of incremental improvements may still be dangerous, especially in the long run. After all, thinking in terms of continuous change in one direction implies that the structure of an industry fifty years from today, for example, will essentially be the same as it is now, with the only difference being that it would use much larger or more efficient technology. A problem arises, however, because the incremental strategy inhibits planners from considering novel technological systems, unusual needs, or a new political-economic environment that might force dramatic changes. Certainly the incremental approach was taken by utility managers for the five decades before the 1980s. After the industry had been established upon firm principles of management, finance, regulation, and technology, managers made incremental changes when necessary for incorporating new equipment and greater usage by customers. But little else changed. Managing along incremental lines was therefore a conservative approach that did not require imaginative thinking about how different the world of the future might be; it simply extrapolated past trends into the decades ahead. The continued use of the incremental strategy helps explain why utility managers did not anticipate the needs and demands of their customers or regulators when stasis occurred. Nor did it help them anticipate and deal with an industry that is currently in the throes of deregulation and restructuring.

To be sure, managers – especially the vast majority with engineering training – wanted to believe that the technology would improve along previous lines. In fact, most managers with whom I talked retained the view that power technology is improving (especially nuclear technology) and that the public and regulatory bodies make it difficult for utility companies to use it. Even if this point is valid, managers should be much more critical of the view of progress in general. Deeply seated in American culture and the engineering mentality, technological "progress" no longer means the same in today's environment as it did twenty or thirty years ago. In the days when the views of the public and engineer-managers coincided, progress meant the same thing to everyone. Today, technological advances are viewed with skepticism by a public enraged by higher costs, an endangered environment, and the presumed abuses of a technical elite. Managers should therefore avoid the impulse to do what engineers do best – namely, attack the problems of the world with technological solutions. In the electric power industry as well as others, managers should therefore avoid the "easy" answer of the technological "fix" – certainly until technologies can be shown to address more effectively to the concerns of all their constituents.

Perhaps engineer-managers can begin blazing this pathway by redefining their concept of "progress." For the first six decades of the twentieth century, progress meant employing improved technology by an increasingly consolidated and centralized core of companies. "Improvements" could be signified by growing thermal efficiency and increases in unit size. Since the 1960s and 1970s, however, these determinants have shown no improvement, suggesting that progress has stopped. But managers need to give up their decades-old mind-set toward progress and realize that it may not have halted altogether. Rather, progress may simply be characterized by different performance criteria that reflect better the values of the people who have a stake in power technology. In other words, instead of higher efficiency or larger scale, progress may now be indicated by a graph that shows decreasing time intervals needed to install 100 MW of new capacity. For financially strapped utilities, this would be real progress because they could begin obtaining a return on investment in less time than would be required for a 1,000-MW unit. In a similar way, an urban utility that needs more capacity would be attracted to a technology (such as the fuel cell) that produces minimal air pollution. Progress would then be signified by a chart showing decreasing numbers of pollutant particles per thousand kilowatt-hours produced. Depending on the needs of individual utilities and those of their constituencies, any and all of these features may be as important as traditional determinants of progress.

Redefining the notion of progress should not be considered unusual if one believes that consensus must be created within a regulated industry. More unusual, perhaps, has been the electric utility industry's ability to maintain a consensus based on one definition of progress for such a long time. But the values of people change, and because the concept of progress has always been value-laden, meaning different things to different groups, its redefinition is just the natural result of people creating and using technology within a system that permits free expression of rights and preferences. While some may criticize this system as being too open, hindering progress in any form, it is nevertheless the type of system that has developed within a pluralistic American society that denies absolute power to a technical elite. It is a system in which utility managers – and the rest of us – must live.

Even so, this first assumption of continued progress will be difficult for managers to overcome because it conflicts with their view that many problems – including social problems – can be solved with technological fixes. A more promising beginning has already been made toward invalidating the second major assumption, namely, the belief in the ability of manufacturers to deliver technologies that would work as promised. Throughout the industry's history, manufacturers performed the research and development, and the users benefited. What could have been better? But when manufacturers began producing unreliable equipment in the late 1960s and 1970s, utilities became less trusting.[4] Fortunately, the industry in 1972 created the Electric Power Research Institute, which helped utilities (and manufacturers) by performing industrywide analyses of technical problems and by providing some R&D resources for overcoming them. (The manufacturers were, of course, still primarily the ones who resolved many of the problems inherent in the overextrapolated equipment.) In order to evaluate and recognize problems with technologies, then, wise practice suggests that companies develop a healthy skepticism of their manufacturers' abilities – despite their successful experiences in the past – and retain some in-house capability for analyzing the state of their technology. Stated differently, managers should not always trust the claims of manufacturers when planning business strategy.

Perhaps most difficult for utility managers will be forsaking the assumption of the benefits of growth. Once perceived as being a "good thing," growth in utility systems' sizes and electricity sales no longer benefited all participants when costs increased and when growth meant environmental degradation. While many utilities have abandoned the approach in recent years, some managers remain leery of shifting to a new, capacity-*saving* strategy, "fearing it would lend credence to a slow-growth ethic considered threatening to the utilities' health," in the words of *The Energy Daily* in 1981.[5]

Despite the reluctance to abandon their belief in growth, the new approach offers some advantages, even to those managers who think that new and better technology is "just around the corner." Abandoning the assumption can provide the necessary "breathing spell" for new innovations to accrue. In other words, companies should continue the recent trend toward load management as a means to reduce the need for new technologies that do not yet exist. Certainly, utilities should alert manufacturers and R&D institutions to their needs, but they should not create an overwhelming demand when opportunities have not yet become tested technologies. As seen in the 1960s and early 1970s, hasty orders for new and bigger technology contributed to manufacturing problems using the design-by-extrapolation technique.

Utility managers, then, must give up their old values about growth – at least for the moment – and learn to do well with extant technology. It is a lesson learned the hard way by some companies in the 1970s. Aside from the PG&E, the NEES, and the TVA – highlighted in a previous chapter – New York City's Consolidated Edison Company altered its growth strategy when it experienced severe problems well before most other utilities. Finding that it lacked generating capacity by the late 1960s – even after installing the giant, but problem-ridden Big Allis 1,000-MW generator in 1965 – the company formally discarded the growth maxim in 1971 when the company's new president, Charles Luce, pressed for conservation and load management.[6] The company's new "Save a Watt" slogan reflected the dramatic change (albeit a reluctant change) in business practices.[7]

As with all prescriptions, this near-term strategy of low growth or no growth may not be the ultimate way to solve the industry's ills. For Consolidated Edison and other companies later in the 1970s and 1980s, the approach seemed attractive at a time when bad economic conditions and public dissatisfaction could not be mitigated by technological progress. So instead of increasing productivity and meeting load projections though the use of *new* technology, as had been the experience of the past, these companies employed load-management techniques to optimize the use of *existent* technology and reduce the need to build new plants. It constituted a strategy that appeared to make sense in an era of technological stagnation.[8] Of course, a new orthodoxy that embraces a capital-minimization strategy of little construction may lead to the creation of management assumptions on the same order as those prevailing earlier. After all, even with conservation and low growth rates, new power capacity will ultimately be needed. Will there be enough alternative energy systems (or even conventional ones) in place to satisfy these needs?[9] As argued in the previous chapter, the answer to this question may require a complete rerationalization of the industry.

An understanding of one's history and the avoidance of management assumptions acquired through years of "progress" will not guarantee success to utility managers. After all, even when managers become conscious of all that can go wrong, external events such as OPEC oil embargoes can still create havoc.[10] Still, by examining historic assumptions and beliefs, managers can become sensitive to business possibilities and potential pitfalls that cannot be anticipated when viewing the world from a jaded perspective. For example, managers can reexamine old assumptions and learn how their previous strategies may now conflict with the values of their constituencies. They can also learn that, no matter how successful ideas had been in the past, management notions cannot be allowed to follow technology in a deterministic fashion. Instead, strategies must direct the use of technologies – and not vice versa – with managers considering the technical and social consequences of each incremental step. Finally, executives can use historical understanding to provide long-term vision that can help them determine whether new situations pose novel challenges that require fundamental changes in business practices.

In dealing with technological stasis and other problems that confront the electric utility industry, then, managers must comprehend and deal with a gamut of new considerations. Perhaps most importantly, they must become accustomed to routine questioning of the values, assumptions, and standard business practices that have accumulated during a long and successful history and that have affected decisions about the future. Managers in this industry – and others as well – should therefore respect their historical background and consider it as another "factor" in creating business policy. Otherwise, in addition to all the other possible pitfalls that can beset an industry, a history of success may sow the seeds for future failures.

Appendix A
Models of technological progress and "stasis"

This book uses the notion of technological "stasis" as a major analytical device. Because of its novelty, this appendix will examine it further within the context of theoretical discussions concerning technological evolution.

Technological progress in the electric power industry resembled advances made in several other industries. In the aviation, petroleum refining, automobile, telecommunications, semiconductor, and scores of other industries, exponential improvement in one or more performance criteria appeared prominently (such as speed, output, cost per unit message, and component density or reliability). But the prosperity that stemmed from technological advances usually did not last forever. In attempts to understand the declining improvement in these technology-based industries, economists and other social scientists have introduced models that employ the concept of technological "maturity." After reaching this stage in "life," a technology advances in small incremental steps and is sometimes superseded by a revolutionary new technology. Productivity improvements then occur rapidly until the industry standardizes its approaches, leading again to a mature technology.

Such a view may be useful for gaining a superficial appreciation of technological "life cycles," but it does not always offer a satisfactory description of the more subtle interplay between economics, engineering, society, and management. In particular, it does not describe well what occurs as incremental advances end in a mature technology nor why, in many cases, people see a need for making revolutionary advances. This book introduced the notion of stasis as a means to help in understanding how and why the electric power industry reached apparent limits to technological improvement.

Models of technological progress

The evolutionary, rather than revolutionary, character of technological innovation in the electric power industry was fundamental to the supply of cheap and abundant electricity. After the boiler, turbine, and generator became standard elements of power-supply technology at the beginning of the twentieth century, improvements consisted mainly of enlarging the size of components and making them more efficient in small complementary and gradual steps: boilers "grew" larger and burned pulverized coal for better combustion at higher temperatures and pressures; turbine blades "stretched" to wrench more power from steam; and generators provided increased electrical output for a specific size of machine by being cooled first with various gases and then liquids. With basic design principles being established early on, advances occurred as engineers gained experience using stronger materials and incorporating innovations from other fields, such as computer-automated control systems. Improvements, then, came from eking out the last percentage point of fuel efficiency, component reliability, or scale economy and by refining systems already in operation. By no means were these trivial tasks.

In the view of model-making historians such as Edward Constant, the pattern of innovation in the electric power industry can be characterized as the "normal" mode of technological evolution. Extending Thomas Kuhn's view of normal activities in science,[1] this type of work consisted of the gradual improvement of basic hardware (the boiler, turbine, and generator system) through procedures and methods well known to the community of practitioners. Normal innovation, which consumes most of the time of practicing engineers, differs markedly from "revolutionary" innovation, however. This type of change consists of the development of a new technology that is accepted because it overcomes limits of performance that were seen to, or are presumed to, exist.[2] In Constant's example, the aviation industry reached a limit in the maximum speed attained by propeller-driven aircraft. "Functional failure" of aircraft trying to breach this barrier had not yet occurred. But because the engineers Frank Whittle and Hans Von Ohain recognized that such a limit would exist – well before others discovered it in practice – they began work on successful turbojet propulsion systems as an alternative. More recently in the microelectronics industry,

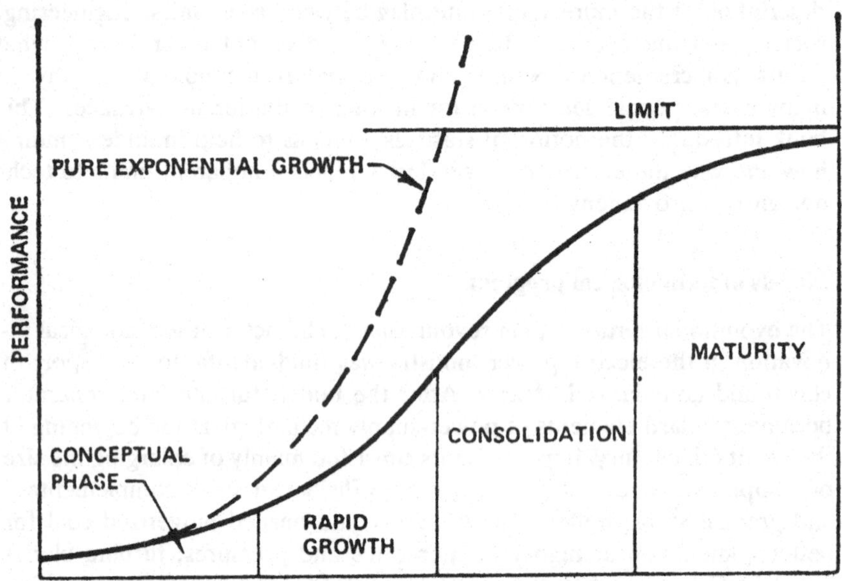

Figure 31. The "S-Curve." The curve suggests the stages of technological evolution: from the conceptual beginnings of a technology to its maturity. Reprinted with permission from the American Association for the Advancement of Science, as appearing in Chauncey Starr and Richard Rudman, "Parameters of Technological Growth," *Science* 182 (October 26, 1973): 360. © 1973 by the AAAs.

physical limits have been overcome before they became a major problem as engineers devised novel methods to squeeze circuits on small chips.[3]

The notion of technological revolutions fits in nicely with the commonly held view (in business circles at least) of the technological life cycle. It is attractive because it offers a model for the development of technology through easily identifiable stages that correlate broadly with the biological life cycle. Derived largely from experience in marketing and international business[4] and applied to technological *products* (versus processes), it identifies a first stage as conceptual birth: when a technology emerges from the research laboratory, tinkerer's garage, or anywhere else, and competes with already existing technologies that provide the same service or product. Then comes rapid growth – as with an infant – followed by consolidation or adolescence, when advances in the technology make it a "winner" in the marketplace due to its superiority over others. This is a period of exponential growth in the technology's use. Finally, as in biological life itself, comes technological maturity, when advances occur – but at a much slower pace.[5]

Graphically, the life cycle has been characterized by an "S-curve" (Figure 31). Knowledge of where a technology lies on the curve is important for business managers. If in the first stage, the technology requires good "breeding" in terms of resources for further development so that it can move to the next, more profitable stage. Competition from other new technologies may be severe in this stage, as they may try to achieve supremacy in the market. If the technology reaches the stage of rapid growth, it requires different marketing and production skills. Most importantly, perhaps, the manager must make sure that rapid innovation continues and that benefits from lower production costs accrue to the company. If the technology ends up in a phase of maturity, however, still other actions are mandated. One strategy may be to induce a technological "revolution" – in Constant's sense – and create a new product that substitutes for and performs better than the previous product. Research-and-development resources would be necessary to do this, and one company or many may compete in the area. Alternatively, if no competing technology exists, a firm may milk this stable, money-making "cash cow" for all its worth, allocating little for improvements. In such a case, fixed costs have been recovered, usually with ample profits, and little supervision has to occur.

In another, more sophisticated model of technological development, changes in product technology are seen as the driving force in the formation of industries. As argued by authors William Abernathy, Kim Clark, and Alan Kantrow in their book, *Industrial Renaissance*,[6] radical innovations in early phases of a technology's development are made in a fluid environment in which the manufacturer attempts to determine the optimum design for the market. But as this design becomes established and as standardization of production techniques increases during the mature phase, innovation in the product becomes less radical and more incremental.

When extended into the realm of technological production *processes*, Abernathy, Clark, and Kantrow have modified the life-cycle model. The new model suggests that production systems "evolve from open-ended and unstructured processes toward rigid and elaborately structured processes."[7] The evolution occurs as products of technological processes themselves change, from unstandardized, low-volume items to mass-produced, standard units. High-volume production often leads to economies of scale when possible and other efforts to obtain further marginal cost reductions as competition firms up. Having committed themselves to this ultimate form of standardization in production processes, managers remain attached to the system "to the bitter end."[8] Often, the bitter end is what is considered "maturity" in an industry, and other companies that can produce a better product or one with new, desirable features take off where the previous one failed. Henry Ford's strategy of perfecting the Model T assembly lines provides a good example of this type of development. By the time the auto producer had reduced costs to an extreme through mechanization of a process technology – leading to a standard product – other companies had introduced innovations in the final product that could not be incorporated into Ford's rigid system. As a result, Ford lost his grip on the automobile market.[9]

While often useful, these interpretations generally presume that product and process evolution moves in one direction only. But even Abernathy, Clark, and Kantrow criticize these traditional evolutionary models because they lead to the conclusion that all industries will inevitably reach maturity and the stage beyond it – death. The conclusion has severe ramifications for the business world: when manufacturers and policy planners believe that maturity is simply a "fact of life," they abandon efforts to revitalize industries and often yield to cries for protectionism.[10] Giving more credit to managers as people who can alter the course of technology, however, Abernathy, Clark, and Kantrow argue that the evolutionary process of innovation can be reversed in bursts of "de-maturity."

To understand this possibility, the authors introduce the notion of the "design concept" as a particular approach to a product's functional requirements. These requirements result from a series of technological choices concerning the composition of a product (for example, will it be made of aluminum or plastic?), its size, and other parameters.[11] While these concepts change along with the marketplace, there may be a single "core concept" that does not change and that may determine the nature of other concepts. For example, in the development of the automobile engine, the core concept consisted of the internal combustion engine rather than the electric motor. Supporting technologies evolved accordingly, favoring carburetors and fuel pumps rather than long-lived chemical batteries for power sources.[12] In many cases, there may be several core concepts, subordinated by a more general "dominant design" that relates the whole technology to its component parts. Through the establishment of hierarchies of core concepts within the framework of a dominant design, innovative activity that once was scattered becomes more focused on secondary concepts. As core concepts stabilize, standardization becomes greater, and competitive advantage becomes more difficult through product innovation.

Within this framework, Abernathy, Clark, and Kantrow discuss maturity in a manufacturing industry in terms of the nature of its technology. This approach differs from others that focus on sales growth or segmentation in markets. As such,

> [a] mature industry is one in which an earlier uncertainty has been replaced by a stability in core concepts, a stability that permits process technology to be embodied in capital equipment or in engineering personnel and purchased in the marketplace. By this line of reasoning, the fundamental characteristics of a mature industry are the stability of its technology and the ease with which it can be copied.[13]

In short, an industry becomes mature as competitive forces drive manufacturers toward standardization. This move "closes down" the amount of technical experimentation in a product, yielding a fixed or closed product that becomes competitively neutral.[14]

These concepts can be extended to describe as mature the process technology used in the electric power industry. The core concept of the steam turbine generator had become established by the twentieth century's second decade, and the small group of manufacturers (with the market lending its hand) settled on standard components (the boiler, turbine, and generator) as the primary ones for power production. Instead of searching for ways to replace the turbine generator, research and development focuses on improving its performance through design changes in turbine blades, generator cooling, and high-temperature-resistant metals, for example.[15] Though dramatic over the long run, advances in power technology did not seem remarkable at any single moment. To the authors of *Industrial Renaissance*, the industry would appear "mature" because "innovations at the mature stage of productive-unit evolution are often virtually invisible to anyone save the engineers actually working on the projects."[16] In short, maturity – though certainly not decay – had set in.[17]

Not included in these descriptions of a mature industry are characteristics that I describe as more sociological and cultural in nature. In today's "high-tech" computer and semiconductor industries, for example (and as was seen in the turn-of-the-century electric power industry), dynamic industries appear daring and exciting to the public. They provide new products and services that capture the imagination. Newspapers and magazines frequently carry stories about technological advances in these industries, and the public makes best sellers out of books chronicling their heroes. (Such was the case with Tracy Kidder's award-winning book about the Data General computer company, *Soul of a New Machine*.) But besides the public, the technical community also provides an indicator of exciting, growing industries by the large number of graduating engineering students that enter these industries. And the business world "votes" as well by offering the highest executive salaries to people in the challenging and growing industries. (See Chapter 9.) In all these "sociological" categories, the electric power industry had left the stage of being a dynamic growth industry by the 1930s. Again, it had become mature.

The notion of technological stasis

I introduced the notion of technological stasis to clarify some of the causes and consequences of "maturity" as used in the previous sense. As defined here, stasis is what appears to be the end of technical advance in an industrial process technology. It is characterized by barriers to further improvement in performance criteria that traditionally marked "progress" in the eyes of the engineers who manufactured and used the technology. In the electric power industry, technological stasis occurred in the 1960s and 1970s, when the most important indicators of advance – thermal efficiency and economies of large scale – no longer showed improvements.

But technological stasis is not just a hardware problem. It is a *systems* phenomenon that comprehends technical and *social* components. The system is one in which hardware develops as a result of interactions between business managers in the vendor and buyer communities, regulatory authorities, financial market makers, and the general public (or consumers). Each participant has a set of values, and when all players develop a consensus, the technology reflects those values. As noted by the historian of technology, John Staudenmaier, "[e]very technology is a human artifact, an artificial construction whose design reflects a limited set of prior technical restraints and a limited set of values within a particular world view."[18] But when the values of the participants change, they can affect the way technological systems emerge and operate. In this history of the power industry, we have seen how several of the stakeholders (i.e., consumers and regulatory bodies) exerted little influence on the way the technology was designed or used before 1970, since their values harmonized with the other constituencies. During the tempestuous 1970s, however, the previously quiescent actors developed values that conflicted with the utilities and their manufacturers, thus creating an environment in which "progress" attained different meanings.

This "multifactor" approach to understanding technological limits encompasses traditional explanations. For example, a technology may reach some limits for relatively simple economic reasons. Such an occurrence might result when the price of a product declines so much in world markets – such as in the aluminum industry – that investment in incremental or revolutionary advances in the process technology is not financially wise. A similar outcome may result if there is already overcapacity in an industry, as in petroleum refining in the mid-1980s. Improved technology may offer productivity enhancements, but it would also make obsolete the heavy capital investment in a previous technology.

But the cause of problems can be more complex. From an engineering point of view, they can arise when hardware systems reach purely physical limits – such as the sound barrier for propeller-driven aircraft. In another example, the limits of mechanical relays in early computers made it impossible to attain the speed of calculations required by scientists. Or, as has been seen in this book, barriers were reached in the strength of standard metals used in power technologies. These limits frequently elude discussion in economic models of technological development.

Additionally, a technology may reach limits because of social or political constraints. In the United States, for example, legislative action and public outcries in the 1970s effectively limited the development of supersonic commercial aircraft and technology for manned space exploration. To be sure, these technologies have never been forced to pass any economic tests, being heavily subsidized by governments. They therefore did not reach limits in the classical sense of confronting economic constraints or physical barriers. Research and development in these areas continued elsewhere, demonstrating that nontechnical "forces" were at work in constraining the technologies in the United States. Models of evolutionary development, however, generally do not comprehend this.[19]

Perhaps most importantly for this study, however, the onset of stasis may be caused by managers of technological systems. Through the use of time-tested (though not reexamined) assumptions and business practices, managers may push for the development of technologies faster than manufacturers would prefer. In response, the vendors may institute design techniques that yield unreliable and poorly performing equipment. These actions may lead to functional failures and what may at first blush appear like the end of technological advance.

Unlike some notions of maturity, "stasis" does not imply that a technology remains in that state forever. Incremental or even revolutionary changes may still occur on some levels, leading to cost savings or improvements among different performance criteria (such as pollution control). Alternatively, new technologies (such as nuclear breeder reactors) may be developed by some players that may not win widespread acceptance due to the actions of others. In short, stasis can be viewed as a dynamic condition – a kind of unstable equilibrium – where technical, managerial, and social "forces" reach a stalemate characterized by unimproving technical performance. Continuing with this analogy from the physical sciences, if one of these forces diminishes (such as the managerial push for more rapid scale increases), technological advances along traditional or novel lines can occur again.

The same point may be made using another analogy. One administrator of power technology research and development suggested that stasis is the equivalent of "constipation."[20] Movement is slow and stubbornly resisted, but possible to surmount with proper treatment. This description is perfectly appropriate. Because stasis arises from a combination of physical, managerial, and social causes, it may be overcome when conditions relating to the causes change. If competition with a standard technology arises, for example, the owners of the old technology might feel the economic incentive to reinvest resources in it, hoping that it can remain competitive. Such is the case for "Indian summer" technologies – those that have fared well for a long time but, under the stimulus of new competition, become advanced well beyond what was previously conceived as possible. This type of rejuvenation occurred in the gas lighting industry in the 1890s. Soon after the introduction of the electric light, the newly invented Welsbach "mantle" provided the means for producing more than twice the illumination from a unit of gas. The old industry had therefore improved its standard technology such that people would be less inclined to opt for the new, expensive – though attractive – electric light.[21] The new mantle extended the industry's life until the electric light advanced sufficiently (and power costs declined enough) to supplant gas lighting. Meanwhile, if the American populace should eventually reach a political consensus concerning the need for and safety of nuclear power, manufacturers will recommence development of the technology.[22] It may then start back up on the standard "S-curve" through gradual improvements, or it could jump to a revolutionary technology.

To conclude, the notion of stasis can accommodate either incremental or revolutionary advances in technologies that appear to have plateaued. Compared to concepts that treat developing technology as a biological system, stasis offers a more flexible framework for analyzing whether a technology deserves further investment of resources. It does this by focusing attention on nontechnical circumstances that contribute to industrial stagnation. Because of this different focus, an understanding of stasis may therefore stimulate actions that could reinvigorate a technology and the industry that uses it.

Appendix B
The efficiency of electric power plants

As noted in the body of this book, the improvement of thermal efficiency of power plants constituted one of the two important trends in the industry's history. Getting more electricity out of a fixed quantity of fuel meant that the industry could reduce its costs (and prices to customers) and increase its productivity. It was also one element of the grow-and-build strategy. But in the 1960s, thermal efficiency no longer improved, with theoretical limits constituting one reason for the plateau. While the text describes some of the ways that engineers improved thermal efficiency and also why thermal efficiency hit barriers, this discussion will offer more details concerning thermodynamic theory.

Efficiency of individual components in a plant

The efficiency of each component within the system that turns heat into electricity combines to yield the overall efficiency of a power plant. In a 1970s-vintage plant, as represented in Figure 32,[1] the boiler converts about 87% of the heat energy from fossil fuel into kinetic energy in steam, which is then passed through a turbine. According to the second law of thermody-

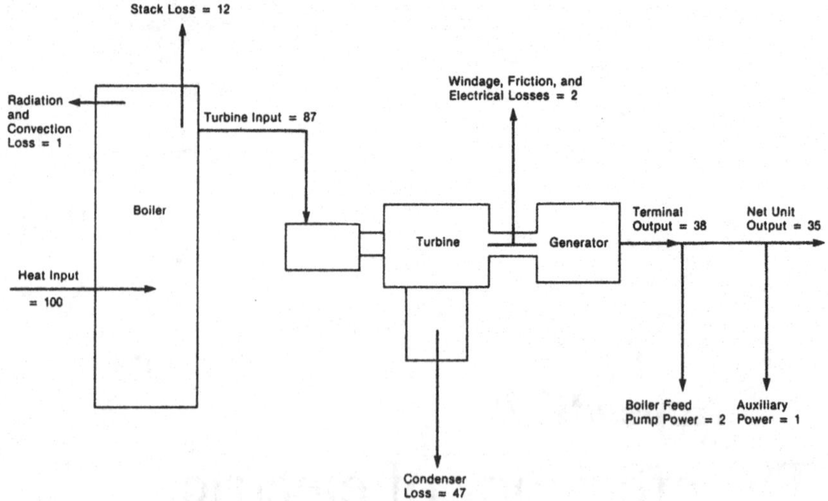

Figure 32. Illustration showing how a portion of heat energy is converted into electricity in a modern (ca. 1975) fossil-fueled power plant. Redrawn diagram based on an illustration in W. D. Marsh, *Economics of Electric Utility Power Generation* (Oxford: Clarendon Press, 1980), p. 101.

namics (see what follows), heat energy must be expelled in order to perform useful work. In the case of this typical plant, 47% of the original energy is expelled into the condenser. At this point, however, the plant has converted heat energy into mechanical motion in the form of a rotating shaft attached to the generator.

The generator constitutes an inherently efficient machine, converting well over 90% of the mechanical energy into electricity. After subtracting for minor losses, 38% of the original energy remains as electricity. But not all the electricity is sent out on transmission lines; some of it is used to pump water through the boiler. More electricity powers auxiliary systems such as antipollution devices. Ultimately, about 35% of the original energy content of the fuel is delivered to the "busbar" – the physical connection between the plant and the transmission system. This, then, is the thermal efficiency of the plant.

Technical achievements for attaining higher efficiency

Higher thermal efficiency resulted through the twentieth century largely from using steam at ever-increasing temperatures and pressures and by reducing heat losses. Aside from the techniques already described in the book, engineers at manufacturing firms also used the technique known as "reheat." This process consisted of bleeding off steam as it passed through the turbine (after it has lost some of its potential energy) and heating it again in the boiler. Though developed before World War II, the process did not prove economical until steam pressures rose beyond 1,450 pounds per square inch (psi) and until temperatures exceeded 900 °F. Once these steam conditions were met in the early 1950s, however, opportunities for reheat beckoned. When first used, the promise of reheat did not disappoint designers: gains of about 5.5% to thermal efficiency resulted almost immediately. The reheat cycle became so popular that by 1954, 92% of General Electric's turbine sales (by total kilowatt shipments) consisted of reheat turbines. Additional reheat stages, in which steam recycled a second time, added another 0.75% improvement at temperatures of 1,050 °F, but further reheat stages demonstrated diminishing returns.[2]

Efficiency also increased as unit scale increased. Here a simple physical (or geometrical) factor comes into play. Because some steam always leaks around turbine blades in the unit casing, energy in the steam is lost and does no useful work. The clearance through the "running seals" in the casing remains the same for both large and small units, so the ratio of leakage to total steam flow path – an indicator of the thermal loss – is smaller for a bigger unit. Size also plays another role. Efficiency is reduced by irregularities in turbine blades, which cause a disturbance in steam flow. For blades that have the same metallic finish, a "pit" on the surface would have a smaller relative effect on a large blade, again making a larger unit more efficient.[3]

The TVA constituted one utility (among many) that found a correlation between increased size of units and thermal efficiency. Jumping from units of 125-MW capacity to those of 275-MW, the agency found efficiency increasing from 36.7 to 37.5%. Even after excluding the effects of increases in steam temperatures and output, TVA engineers noted a residual gain of 0.23% attributable primarily to size. They also expected a 0.9% improvement to result when they moved up from units that produced 500 to 900 MW.[4] More general experience from 184 units installed in the United States between 1968 and 1972 demonstrated that most of the improvement in thermal efficiency occurred in large-sized units as manufacturers improved several aspects of the technology in the same machine.[5]

Limits to thermodynamic efficiency

The second law of thermodynamics governs the operation of all "ideal" engines that employ heat to make mechanical motion or to perform any other form of work. Unlike mechanical systems that can attain close to 100% efficiency, heat systems are constrained by the second law of thermodynamics, which requires that a large proportion of the heat supplied to the system at a high temperature be rejected to a low-temperature receiver (called the condenser) in order to produce power continuously. This maximum possible theoretical efficiency for assumed high and low temperatures is usually called the "Carnot-cycle efficiency," named after the French physicist who first outlined the theory in the early nineteenth century. It can be calculated for an ideal system that uses a "perfect," noncondensable gas using the equation:

$$\text{Efficiency} = 1 - T_l/T_h$$

where T_h is the higher temperature, and T_l is the lower, condenser temperature, both measured in absolute temperatures.

The Carnot-efficiency law provides an upper limit for the thermal efficiency of a power plant. Given the practical situation of a condenser operating around 60 °F (about 520 degrees Rankine – an absolute temperature scale[6]) and an extreme upper temperature limit of about 2,540 °F (about 3,000 °R) for an engine burning carbon fuels, the efficiency becomes 83%. If we further limit the highest temperature to 1,540 °F (2,000 °R), which constitutes the maximum temperature that can be maintained for any length of time with metals currently available, the maximum efficiency for a machine – no matter how cleverly designed – becomes 74%.[7]

Unlike the Carnot cycle, which describes ideal engines, the Rankine cycle (named after the nineteenth-century Scottish physicist, William J. M. Rankine) applies to "real" engines, with efficiency calculated on the basis of the temperature and pressure of steam – not just temperature. As noted in Chapter 7, Rankine-cycle efficiency began reaching the point of diminishing returns in the 1960s. Even if temperatures and pressures could be increased without metallurgical failures, the payoff in improvements in thermal efficiency gradually became smaller, as is suggested by the table. It demonstrates that increases in steam pressure (in 1,000-psi increments) produces ever-smaller decrements in heat rate (or increments of thermal-efficiency improvements). As pointed out in the otherwise optimistic *National Power*

Improvement in theoretical cycle heat rate
(Btu/kWh per 1,000-psi increase in pressure)

Pressure from	Change to	Non-reheat	Single Reheat	Double Reheat
Based on initial and reheat temperatures of 1,000 °F				
2,000	3,000	305	265	245
3,000	4,000	185	160	150
4,000	5,000	135	125	115
5,000	10,000	67	64	60
Based on initial and reheat temperatures of 1,600 °F				
2,000	3,000	235	185	180
3,000	4,000	155	125	115
4,000	5,000	105	100	80
5,000	10,000	68	54	50

Survey of 1964, the reason for the reduced improvements was that "even on a purely theoretical basis, there is much less to be gained by increasing pressure from 3,500 to 10,000 pounds per square inch than from 1,200 to 3,500 pounds per square inch."[8] In other words, the Rankine cycle and realities concerning metallurgical failure at high temperature and pressure also limited thermal efficiencies to below 50% when using conventional fossil-fuel systems.

Bibliographic note

Much of the information used in this book depended on the reminiscences and insights offered by utility engineers, managers, and regulators. Speaking to me ten or more years after the events they described, these people tended to be more reflective and amenable to discussions of events that occurred "in the distant past" than about recent utility policies – policies that have not yet been tested by time. Listed are the names and affiliations (at the time of the interviews) of people who talked with me. I am most grateful for their assistance.

Consolidated Edison Company, New York, NY.

Louis Sandhop, manager, Community Relations, July 14, 1982.
Arthur W. Flynn, Technical Consultant, July 14, 1982.
Two other managers from the company who prefer to remain unmentioned.

Electric Power Research Institute, Palo Alto, CA.

Thomas H. McClosky, Department of Availability and Life Extension, July 12, 1982.

David M. Rigney, project manager, Fuel Cell Applications, telephone interview, February 7, 1984.

Ed Gillis, program manager, Fuel Cells and Hydrogen Technology program, telephone interview, December 9, 1983.

Anthony F. Armor, program manager, Performance and Advanced Technology, July 12, 1982.

Floyd L. Culler, president, January 16, 1986 and January 17, 1986.

Chauncey Starr, vice chairman, January 16, 1986.

Sam H. Schurr, deputy director, Energy Study Center, January 16, 1986 and January 17, 1986.

Richard Balzhiser, senior vice president, Research and Development, January 16, 1986 and January 17, 1986.

Walt Esselman, director, Engineering Assessment and Analyses, Research and Development, January 16, 1986.

Richard Zeren, director, Planning and Evaluation Division, January 16, 1986.

Stanley Vejtasa, manager, Technology Evaluation, Planning and Evaluation Division, January 16, 1986.

Katherine A. Miller, senior planning economist, Planning and Evaluation Division, January 16, 1986.

Oliver S. Yu, manager, Planning Analysis, Planning and Evaluation Division, January 16, 1986.

Fritz Kalhammer, vice president and director, Energy Management and Utilization, January 16, 1986.

Edwin Zebroski, chief nuclear scientist, Energy Study Center, January 16, 1986.

Richard Rudman, vice president, Industry Relations and Information Services Group, January 17, 1986.

George T. Preston, director, Fossil Fuel Power Plant, January 17, 1986.

Rene Males, vice president, Energy Analysis and Environment, January 17, 1986.

John J. Dougherty, vice president, Electrical Systems Division, January 17, 1986.

Arnold Fickett, department director, Energy Utilization, January 17, 1986.

Kurt Yeager, vice president, Coal Combustion Systems Division, January 17, 1986.

Ronald H. Wolk, director, Advanced Fossil Power Systems Department, January 17, 1986.

John Cummings, department director, Renewable Resources Division, January 17, 1986.

Bechtel Power Corporation, San Francisco, CA.

H. Gene Hally, principal engineer, San Francisco Power Division, July 9, 1982.

Pacific Gas and Electric Company, San Francisco, CA.

Peter Hindley, senior mechanical engineer, Generation Planning Department, July 8, 1982.
Greg S. Pruett, Public Information representative, July 8, 1982.
Chris Piper, News Bureau, July 8, 1982.

Babcock and Wilcox, Barberton, OH.

Neil Eft, sales promotion engineer, telephone interview, June 19, 1984.

Appalachian Power Company, Roanoke, VA.

Barry L. Thomas, supervisor, Rate Design and Research, Rates and Contracts Department, September 21, 1981.

Virginia Electric and Power Company, Richmond, VA.

Samuel C. Brown, senior vice president, Power Station and Construction, August 20, 1981.
C. M. Jarvis, vice president, August 19, 1981.
Samuel A. Hall, III, director, Rate Application, August 19, 1981.
H. M. (Mack) Wilson, manager, Rates, August 19, 1981.
Jerry Causey, marketing manager, August 19, 1981.
E. Paul Hilton, director, Rate Design, August 19, 1981.
H. H. Dunstan, Jr., manager, Cost Allocation Studies, August 19, 1981.
Irene M. Moszer, manager, Forecasting and Economic Analysis, August 20, 1981.

General Electric Company, Schenectady, NY.

Louis Coffin, internal GE consultant with specialty in mechanics of materials, telephone interview November 12, 1982.
George B. Cox, senior vice president and group executive, Turbine Business Group, May 21, 1984.
Charles Elston, retired manager of large steam turbine business, telephone interview, November 19, 1982.
Walter L. Marshall, manager, Turbine Technology Laboratory, May 22, 1984.
Norman H. Jones, manager, Generator Engineering Section, May 23, 1984.
Randall J. Alkema, general manager, Large Steam Turbine-Generator Department, May 21, 1984.
Robert S. Couchman, manager, Turbine Engineering Section, May 22, 1984.

Hugh J. Murphy, general manager, Large Steam Turbine-Generator Marketing Operations, May 21, 1984.

Richard Alben, consultant, Technology Assessment, Engineering Physics Laboratory, May 23, 1984.

George Wise, historian, GE Corporate Research Laboratories, May 23, 1984.

Westinghouse Electric Corporation, Pittsburgh, PA, and Orlando, FL.

Samuel Herwald, retired WE manager, Herwald Consulting Services, telephone interview, November 10, 1982.

James A. Thurber, former human relations manager, April 7, 1986. (His position at time of the interview was as director, Human Resources, New England Medical Center.)

Ralph Young, project manager, April 24, 1986.

J. David Conrad, manager, Business Development, November 13, 1985; August 9, 1985; and April 24, 1986.

Alan D. Cooper, manager, Power Generation Quality Assurance, November 13, 1985.

George Dann, manager, Human Resources, November 13, 1985.

Eugene J. Cattabiani, executive vice president, Power Generation, July 29, 1985.

Theodore Stern, executive vice president, Nuclear Energy Systems, July 29, 1985.

John F. Traexler, manager, Engineering, Steam Turbine-Generator Division, April 25, 1986.

Lon W. Montgomery, manager, Electrical Stator Systems Engineering, several letters.

Tennessee Valley Authority, Chattanooga and Knoxville, TN.

S. David Freeman, director, January 17, 1984.

Hugh G. Parris, manager of Power, January 19, 1984.

E. Floyd Thomas, retired manager of Power Operations, January 19, 1984.

G. O. Wessenauer, retired manager of Power, January 19, 1984.

Aubrey J. "Red" Wagner, former chairman of the board, January 18, 1984.

George Palo, former manager of Engineering Design and Construction, January 17, 1984.

Dwight Patterson, assistant to the manager of Engineering Design, January 18, 1984.

James L. Williams, director of Purchasing, January 19, 1984.

George H. Kimmons, manager of Engineering Design and Construction, January 17, 1984.

Carl Crawford, manager of Power Information, January 19, 1984.

C. John Morris, chief, Visitor Relations Staff, January 17, 1984.

Boston Edison Company, Boston, MA.

James M. Lydon, executive vice president, May 23, 1985.
Priscilla Korrell, Public Information officer, May 23, 1985.

New England Electric System, Westborough, MA.

Joan T. Bok, chairman of the board, May 8, 1985.
Guy W. Nichols, former chairman of the board, May 8, 1985.
Fred Greenman, legal counsel, May 8, 1985.
Gerald Browne, Planning, September 25, 1985.
Fred Pickle, Alternative Energy Planning, September 25, 1985.
Michael Jesanis, Finance Department, September 25, 1985.
Lydia Forsuckich, Load Forecasting, September 25, 1985.

American Electric Power Service Corporation, Columbus, OH.

John E. Dolan, vice chairman, Engineering and Construction, July 30, 1985.
Dorothy Miesse, former personal secretary to Philip Sporn, July 30, 1985.
William W. Corbitt, vice president, Communications, July 30, 1985.

United Technologies, South Windsor, CN.

Don McVay, marketing representative, Fuel Cells Power Systems, telephone interview, December 14, 1983.

Gas Research Institute, Chicago, IL.

John Cutica, telephone interview, January 13, 1984.

Commonwealth Edison Company, Chicago, IL.

Bide L. Thomas, executive vice president, August 20, 1984.
John J. Viera, vice president, Marketing and Load Management, August 20, 1984.
Robert L. Bolger, assistant vice president, August 20, 1984.
Robert Beckwith, manager, System Planning, August 20, 1984.
Gary R. Grable, director of Advertising, August 20, 1984.

Texas Utilities, Dallas, TX.

Perry G. Brittain, chairman of the board and chief executive officer, January 14, 1985.
Burl B. Hulsey, Jr., vice chairman of the board, January 14, 1985.

Madison Gas and Electric Company, Madison, WI.

Frederick Mackie, former president, general manager, and chairman of the board, August 21, 1984.

Leo Brodzeller, former group vice president and secretary, August 21, 1984.

Gerald Wilson, senior vice president, Administration, August 21, 1984.

Richard M. Lawrence, vice president, Public Affairs, August 21, 1984.

Steve Kraus, Public Affairs representative, August 21, 1984.

Duke Power Company, Charlotte, NC.

S. B. "Pete" Hager, Design Engineering, August 8, 1985.

Robert E. Miller, Principal Engineer, Design Engineering, August 8, 1985.

Dennis Murdock, Electrical Division, August 8, 1985.

Robert F. Edmonds, senior engineer, August 8, 1985.

Toney A. Mathews, Project Management Division, Design Engineering Department, August 8, 1985.

Christopher C. Rolfe, manager, Research and Projects, Design Engineering, August 8, 1985.

Rick B. Priory, vice president, Design Engineering Department, August 8, 1985.

Regulators

Joseph C. Swidler, former FPC chairman and chairman of the New York State Public Service Commission, telephone interview March 15, 1983.

Carl E. Bagge, former FPC Commissioner, currently president, National Coal Association, Washington, D.C., March 11, 1983.

Alfred Kahn, former chairman, New York State Public Service Commission, November 9, 1983.

Notes

Preface

1. Nathan Rosenberg, "Technology," in Glenn Porter, ed., *Encyclopedia of American Economic History*, Vol. 1 (New York: Scribners, 1980), pp. 294–5. Also see Zvi Griliches, "R&D and Productivity: Measurement Issues and Econometric Results," *Science* 237 (July 3, 1987): 31–5.
2. As John Staudenmaier has pointed out, historians have perpetuated a "Whiggish" belief in the equivalence of technology and progress by the attention they pay to success stories. See John M. Staudenmaier, *Technology's Storytellers: Reweaving the Human Fabric* (Cambridge, MA: The MIT Press, 1985).
3. Edward W. Constant II, *The Origins of the Turbojet Revolution* (Baltimore: Johns Hopkins University Press, 1980).
4. The history of the petrochemicals industry after World War II parallels some aspects of the history of the utility industry especially well: for several years, the petrochemicals industry enjoyed exponentially rising demand for products and incremental improvements with large-scale technology. But in the late 1960s and early 1970s, managers recognized that new, large-scale technologies would not solve the industry's problems. As a result, the scale and performance of the industry's technology appeared to falter. See Peter H. Spitz, *Petrochemicals: The Rise of an Industry* (New York: Wiley, 1988). For information on the effects of technologi-

cal improvements and scale economies in the petrochemicals industry, also see John L. Enos, "A Measure of the Rate of Technological Progress in the Petroleum Refining Industry," *Journal of Industrial Economics* 6 (1958): 180–97.

5. Technological advances still are being made in long-distance transmission, as fiber-optics technology replaces (and augments) microwave and satellite transmission systems. Nevertheless, transmission costs represent only about 20 to 25% of the total costs of a long-distance call, with billing, access to long-distance networks, and marketing costs accounting for the balance. Personal communication between Harvard Business School Research Associate Dekkers L. Davidson and the author, November 8, 1984. See also Dekkers L. Davidson, "Telecommunications in Transition," Case 0–384–207, *Harvard Business School Case Studies Series* (1984), pp. 19–20.

6. This quotation refers to David M. Potter's book, *People of Plenty: Economic Abundance and the American Character* (Chicago: University of Chicago Press, 1962).

7. See, for example, William F. Ogburn, "National Policy and Technology," in U.S. National Resources Committee, Science Committee, *Technological Trends and National Policy, Including the Social Implications of New Inventions*, Chapter 1. (Washington, DC: U.S. Government Printing Office, 1937); John Jewkes, David Sawers, and Richard Stillerman, *The Sources of Invention* (London: Macmillan, 1958); Jacob Schmookler, *Invention and Economic Growth* (Cambridge, MA: Harvard University Press, 1966); Edwin Mansfield, *The Economics of Technological Change* (New York: Norton, 1968); and F. M. Scherer, *Innovation and Growth: Schumpeterian Perspectives* (Cambridge, MA: The MIT Press, 1984).

8. These terms are taken from Thomas P. Hughes, "The Electrification of America: The System Builders," *Technology and Culture* 20 (1979): 124–61.

Introduction

1. Among the many books that focus on these factors alone are Leonard S. Hyman, *America's Electric Utilities: Past, Present, and Future* (Arlington, VA: Public Utilities Reports, 1983); and Scott Fenn, *America's Electric Utilities: Under Siege and in Transition* (New York: Praeger, 1984).

2. Between 1961 and 1969, for example, capacity additions in the United States accounted for 41% of the noncommunist world's total. As early entrants in the field, the General Electric Company and Westinghouse Electric Corporation excelled in the American market and then dominated the world market as well. Even with the wave of mergers among European manufacturers in the 1960s, American companies maintained commercial supremacy. Moreover, several European and Japanese manufacturers depended on American know-how through licensing and partial ownership agreements with General Electric and Westinghouse. Excluding the role of these agreements with foreign firms, American manufacturers held almost 32% of the world market for power equipment in 1955, while Great Britain and West Germany garnered 22% and 19%, respectively. By 1969, these figures had readjusted to 23%, 23%, and 13%. This discussion of the role of American manufacturers in the world market draws from A. J. Surrey and J. H. Chesshire, *The World Market for Electric Power Equipment: Rationalisation and Technical Change* (Sussex: University of Sussex Science Policy Research Unit, 1972), pp. 2–10.

3. In nationalized power systems that existed in much of the world, power technology was added in standardized, incremental steps at a slower rate than in the United States. A comparison of French, English, and American utility systems between 1948 to 1963 showed, for example, that in the main technical characteristics of generating equipment (output of turbine generators and boilers, and temperature and pressures of steam), Britain lagged up to four years behind France and up to nine years behind the United States. See F. P. R. Brechling, "An International Comparison of Production Techniques: The Coal-Fired Electricity Generating Industry," *National Institute Economic Review* 36 (May 1966): 30–42. A graph that compares the capacity of power units from 1950 through 1975 in the United States, France, England, and Germany is presented in Von Helmont Scheffczyk, "Dampfturbinen," *Elektrizitaetswirtschaft* 75 (1976): 695. It shows that American utilities typically used units whose output greatly exceeded those used in European countries.

4. To be fair, thermal efficiency of power units is a function of technological design and operational use. A unit that is operated consistently with a steady load will have a higher thermal efficiency. Nevertheless, as will be seen later, technological design improvements appeared to plateau in the 1960s, and operational excellence could not make the power units achieve higher efficiencies than what the basic technology permitted.

Chapter 1

1. The concept of dominant design was first used for describing the development of a commercial technology or manufacturing process in William J. Abernathy, Kim B. Clark, and Alan M. Kantrow, *Industrial Renaissance: Producing a Competitive Future for America* (New York: Basic Books, 1983).

2. See Thomas P. Hughes, "The Electrification of America: The System Builders," *Technology and Culture* 20 (1979): 124–61. Much of the discussion about the early days of the industry derives from this article and Professor Hughes' *Networks of Power: Electrification in Western Society, 1880–1930* (Baltimore: Johns Hopkins University Press, 1983).

3. *Ibid.*, pp. 45–6.

4. See the first several chapters in *ibid.*

5. Among the more colorful battles in the "war of the currents" was Edison's promotion of alternating-current electricity for use in executing prisoners in New York State. Edison hoped that the public would associate Westinghouse's AC system with life-threatening danger, thereby making the DC system more popular. Even though New York State used AC for its first-of-its-kind electric chair in 1890, AC triumphed in the long run. For excellent descriptions of the "war," see Theodore Bernstein, "A Grand Success," *IEEE Spectrum* (February 1973): 54–8; and Terry S. Reynolds and Theodore Bernstein, "The Damnable Alternating Current," *Proceedings of the IEEE* 64 (1976): 1339–43.

6. Hughes, "The Electrification of America," p. 139.

7. "Rates and Regulation," *Electrical World* 181 (June 1, 1974): 358.

8. Hughes, *Networks of Power*, p. 45.

9. W. J. Greene, "A Method of Calculating the Cost of Furnishing Electric Current and a Way of Selling It," *Electrical World* 27 (February 29, 1896): 222–3; in Edison Illuminating Company of Detroit, *The Development of Scientific Rates for Electric Supply* (Detroit: private printing, 1915), pp. 23–4.

10. Arthur Wright, "Cost of Electricity Supply," *Minutes of Municipal Electric Association* (1896), pp. 1–8, in Edison Illuminating Company, *The Development of Scientific Rates*, pp. 33–4.

11. The Association of Edison Illuminating Companies, an organization that Insull had promoted in 1885, developed an explanation for the phenomenon. Because electricity could not be stored economically (as could gas), generating plants had to be built large enough to accommodate the maximum demand of the day (in the evening) and year (in the winter, when daylight hours were few). As a result, constructing the plants and laying the cables was expensive, even though for most of the time, the full capacity of the equipment lay idle. Thus, the fixed costs of interest, maintenance, and depreciation were high, whereas operating costs (mostly fuel) remained relatively low and based only on the volume of sales. Therefore, by adding new customers, the electric companies increased their investment, making it more expensive and less profitable to provide the service. In order to make good money on investments, so this explanation went, companies had to maintain a small number of customers and induce them to use more electricity or charge more for what they already used. Consequently, many managers concluded that lighting would remain forever a luxury product. See Forrest McDonald, *Insull* (Chicago: University of Chicago Press, 1962), pp. 63–4.

12. John Hopkinson, "On the Cost of Electric Supply," *Transactions of the Junior Engineering Society*, 3 Part 1 (1892): 1–14, in Edison Illuminating Company, *The Development of Scientific Rates*, p. 9.

13. *Ibid.*, p. 10.

14. *Ibid.*, p. 15.

15. Wright, "Cost of Electricity Supply," p. 36.

16. *Ibid.*, pp. 39–45.

17. Samuel Insull, "Sell Your Product at a Price Which Will Enable You to Get a Monopoly," in William E. Keily, ed., *Central Station Electricity Service: Its Commercial Development and Economic Significance as Set Forth in the Public Addresses (1897–1914) of Samuel Insull* (Chicago: private printing, 1915), pp. 116–17. Also see James C. Bonbright, *Principles of Public Utility Rates* (New York: Columbia University Press, 1961), p. 309, for an example of how lower rates for "good" industrial customers can ultimately lead to lower rates for residential users.

18. Not everyone applied these principles as effectively as Insull. In the 1930s, the publicly owned Tennessee Valley Authority would reemphasize the same principles in a bold fashion.

19. Raymond C. Miller, *Kilowatts at Work: A History of the Detroit Edison Company* (Detroit: Wayne State University Press, 1957), p. 17.

20. Samuel Insull, "Stepping Stones of Central-Station Development Through Three Decades," lecture delivered before the Brooklyn Edison Company Section of the National Electric Light Association, June 26, 1912, in Keily, *Central Station Electric Service*, p. 353. The largest reciprocating steam engine built for producing electricity was 5,000 kW, installed for the New York City transit system just after 1900. "A Century of Power Progress: Enter Steam Turbines and Diesels," *A Century of Power Progress*, published by the editors of *Power* (New York: McGraw-Hill, 1982), no page, reproduced from *Power* magazine, May 1981.

21. See General Electric, *A Highlight History of General Electric* (Schenectady: General Electric Company, 1978), p. 16; and R. H. Parsons, *The Early Days of the Power Station Industry* (Cambridge: Cambridge University Press, 1939). A popularized history of the Parsons engine is W. Garrett Scaife, "The Parsons Steam Turbine," *Scientific American* 252 (April 1985): 132–9.

22. Insull recounted his negotiations with General Electric in Samuel Insull, "Massing of Energy Production an Economic Necessity," a speech given at a General Electric dinner in Boston, February 25, 1910; in Keily, *Central Station Electric Service*, pp. 137–8.

23. Hughes, "The Electrification of America," pp. 145–6.

24. Charles E. Neil, "Entering the Seventh Decade of Electric Power," *Edison Electric Institute Bulletin* 10 (September 1942): 328–9.

25. Samuel Insull, "The Production and Distribution of Energy," lecture delivered before the Franklin Institute in Philadelphia on March 19, 1913; in Keily, *Central Station Electric Service*, p. 365.

26. Trade journals were full of stories describing the benefits of interconnections after World War I. For example, see F. C. Sargent, "Interconnection is Successful in Eastern Massachusetts," *Electrical World* 78 (July 30, 1921): 216–17. The article points out that interconnection of three small firms saved them several million dollars in investment for new plants and allowed them to reduce coal consumption by 10,000 tons per year. Also see "An Interconnection of Increasing Value," *Electrical World* 78 (July 30, 1921): 204; and "New Southern Interconnection Relieves Power Shortage in the Carolinas," *Electrical World* 78 (October 29, 1921): 890. The notion of interconnecting complete regions of the country to realize these benefits on an even larger scale became popular in the 1920s, often going by the term "superpower." See "Opportunity for Wider Interconnections," *Electrical World* 83 (January 12, 1924): 75; and Herbert Hoover (then Secretary of Commerce), "Superpower and Interconnection," *Electrical World* 83 (May 24, 1924): 1078–80. However, an attempt forwarded by the state government to create a superpower network that would benefit rural communities in Pennsylvania failed to succeed. See Thomas Parke Hughes, "Technology and Public Policy: The Failure of Giant Power," *Proceeding of the IEEE* 64 (1976): 1361–71.

27. Strictly speaking, the marginal cost is the cost of producing one more unit of a product. Also termed "incremental" cost, it is the cost of producing an additional unit of something. In the utility industry, the marginal cost of capacity equals the cost of building the most recent power unit. Meanwhile, the average cost of construction refers to the total cost of all the plants divided by the total amount of capacity. When marginal costs for new plants decline, the average cost will also fall. For example, if the previous 200 MW of a utility's capacity (consisting of 4–50-MW units) cost $40 million (or $200 per kW) and a new 100-MW unit costs $10 million ($100 per kW), the marginal cost per kilowatt is $100, and the average cost of construction would have fallen to $167 per kilowatt. The concept of marginal costs is explained in detail, with applications to the utility industry and other regulated industries, in Alfred E. Kahn, *The Economics of Regulation: Principles and Institutions*, Vol. 1 (New York: Wiley, 1970), pp. 65–6.

28. Hughes, "The Electrification of America," p. 141. Insull enjoyed making public appearances and explaining to managers, engineers, and the general public his "secrets" of success and low-cost electricity. He often compared, for example, the rates his company charged with those in other parts of the country. In 1914, Insull's company received an average of 2.05 cents per kilowatt-hour. Boston utilities received 5.37 cents per kilowatt-hour, while New York companies earned 4.45 cents per kilowatt-hour. Among large cities, only San Francisco could provide cheaper electricity, due to the use of hydroelectric sources, at 1.97 cents per kilowatt-hour. See Samuel Insull, "Centralization of Energy Supply," address delivered before the Finance Forum of the Young Men's Christian Association in New York, April 20, 1914; in Keily, *Central Station Electric Service*, p. 460.

29. Samuel Insull, "Standardization, Cost Control of Rates, and Public Control," in Keily, *Central Station Electric Service*, pp. 34–47.

30. "Municipal Electric Systems," *Electrical World* 131 (May 21, 1949): 15.

31. More discussion of public power systems can be found in Richard Rudolph and Scott Ridley, *Power Struggle: The Hundred-Year War over Electricity* (New York: Harper and Row, 1986), pp. 22–56; and Thomas K. McCraw, *TVA and the Power Fight: 1933–1939* (Philadelphia: Lippincott, 1971), pp. 1–25.

32. McDonald, *Insull*, p. 114.

33. *Ibid*., p. 119.

34. *Ibid*., p. 128.

35. This last standard was established in 1898 with the U.S. Supreme Court's ruling in the *Smyth v. Ames* case concerning fair returns for the Union Pacific Railroad. The origins of public utility regulation are detailed in Ben W. Lewis, "Public Utilities," in Leverett S. Lyon and Victor Abramson, *Government and Economic Life*, Vol. 2 (Washington, DC: Brookings Institution, 1940), pp. 616–745. A more popularized and colorful account can be found in Thomas K. McCraw, *Prophets of Regulation* (Cambridge, MA: Harvard University Press, 1984).

36. In efforts to prevent changes in the way regulatory commissions operated, utility spokesmen sometimes gave them exaggerated credit for reducing the cost of electricity and for building the public's confidence in the integrity of power companies. See "State Regulation Has Proved Itself," *Electrical World* 93 (May 11, 1929): 915; and W. J. Hagenah, "Soundness of Regulation Proved," *Electrical World* 93 (June 8, 1929): 1170.

37. Hughes, "The Electrification of America," p. 154.

38. *Ibid*., pp. 154–5.

39. See a discussion of holding companies in Charles F. Phillips, Jr., *The Regulation of Public Utilities: Theory and Practice* (Arlington, VA: Public Utility Reports, 1984), pp. 526–33.

40. Hughes, "The Electrification of America," pp. 156–7.

41. Carl D. Thompson, *Confessions of the Power Trust* (New York: Dutton, 1932). This book summarizes testimony given in Federal Trade Commission hearings in 1928 concerning "power trust" abuses. It includes discussions of stock "watering," overcapitalization of holding companies, and propaganda methods used to oppose public power efforts and to spread the "good" word about power companies.

42. Kenneth Jameson, "Economies of Scale in the Electric Power Industry," in Kenneth Sayre, ed., *Values in the Electric Power Industry* (Notre Dame, IN: University of Notre Dame Press, 1977), p. 119. For an exposé on the abuses of holding companies written after many of them collapsed during the Depression, see J. V. Garland and Charles F. Phillips, *The Crisis in the Electric Utilities*, Vol. 10, No. 10, in *The Reference Shelf* series (New York: H. W. Wilson, 1936).

43. "Sweeping Changes in Store as the Industry Enters a Second Century," *Electrical World* 181 (June 1, 1974): 47.

44. Jameson, "Economies of Scale," pp. 119–20.

45. "The Light and Power Industry in the United States," *Electrical World* 93 (January 5, 1929): 36–7.

46. For much more detail on the development of power systems in the United States, Germany, and Great Britain up to 1930, see Thomas P. Hughes, *Networks of Power*; and Leslie Hannah, "Public Policy and the Advent of Large-Scale Technology: The Case of Electricity Supply in the U.S.A., Germany, and Britain," in Norbert Horn and Juergen Kocka, eds., *Law and the Formation of the Big Enterprises in the 19th and Early 20th Centuries: Studies in the History of Industrialization in Germany, France, Great Britain, and the United States* (Goettingen: Vandenhoeck and Ruprecht, 1979), pp. 577–89.

Chapter 2

1. James Prescott Joule, "Investigations in Magnetism and Electromagnetism," *Annals of Electricity* 4 (1839); in James Prescott Joule, *The Scientific Papers of James Prescott Joule*, Vol. 1 (London: Dawsons of Pall Mall, 1887; republished 1963), p. 14.
2. Thomas Ewbank, *Report of the Commissioner of Patents*, 1849, cited in Richard B. DuBoff, *Electric Power in American Manufacturing, 1889–1958* (New York: Arno Press, 1979), p. 1.
3. Elihu Thomson, "Future Electrical Development," *New England Magazine* 6, ns. (1892): 623–35.
4. A. N. Brady, "Electricity in the Service of Man," *North American Review* 173 (1901): 677–83.
5. *Ibid.*, p. 678.
6. *Ibid.*, p. 677. For other statements of service and marvel, see Arthur Goodrich, "Providing the World with Power," *World's Work* 7 (February 1904): 4429–34; and Bernard Meiklejohn, "A New Epoch in the Use of Power," *World's Work* 8 (August 1904): 5084–8.
7. Rosalind Williams, "Reindustrialization Past and Present" *Technology Review* 85 (November/December 1982): 48–57.
8. L. Frank Baum, *The Master Key: An Electrical Fairy Tale Founded Upon the Mysteries of Electricity and the Optimism of its Devotees.* (Indianapolis, IN: Bowen-Merrill, 1901). The quotation comes from the subtitle of the book on its title page.
9. Besides this book, youngsters could catch the electric "bug" by reading any of several biographies of Edison – the first one published in 1879, when the inventor was only 32 years old! See J. B. McClure, *Edison and His Inventions* (Chicago: Rhodes and McClure, 1879).
10. John MacMillan Brown (pseud. Godfrey Sweven), *Limanora: The Island of Progress* (New York: Putnam, 1903), cited in Nancy Knight, " 'The New Light': X-Rays and Medical Futurism," in Joseph J. Corn, ed., *Imagining Tomorrow: History, Technology, and the American Future* (Cambridge, MA: The MIT Press, 1986), pp. 29–30
11. Soloman Schindler, *Young West: A Sequel to Edward Bellamy's Celebrated Novel "Looking Backward"* (Boston: Arena, 1894), p. 45, cited in Howard P. Segal, "The Technological Utopians," in Corn, *Imagining Tomorrow*, p. 123.
12. Segal, "Technological Utopians," p. 127.
13. Rosalind Williams, "The Empress Has New Clothes," *Technology Illustrated* 3 (March 1983): 75.
14. *Home Communications* (January 1984): 11, and Ellen Posner, "Learning to Love Ma Bell's New Building," *Wall Street Journal* (October 12, 1983), p. 26. Still

more mythical symbolism can be found on engraved securities certificates of utility companies. For example, the Wisconsin Electric Power Company stock certificate showed a robed Grecian woman sitting on a globe and offering a radiating light bulb to humankind. When renamed the Wisconsin Energy Corporation, the company redesigned its certificates.

15. Interestingly, most commentators early in the twentieth century discounted the role of steam engines in making electricity. They generally expected power to be produced without air pollution from water turbines.

16. F. B. Crocker, "The Electric Distribution of Power in Workshops," *Journal of the Franklin Institute* 151 (January 1901): 8; cited in Warren D. Devine, Jr., "From Shafts to Wires: Historical Perspective on Electrification," *Journal of Economic History* 43 (1983): 364.

17. For discussions of how electrification increased industrial productivity in the early twentieth century, see Lewis Mumford, *Technics and Civilization* (New York: Harcourt, Brace, and World, 1963; originally published in 1934), pp. 224–7. Mumford not only noted how the electric motor transformed the factory, but how electricity could provide "small-scale industry a new lease on life" (p. 225). Because electric motors are efficient even in small sizes, some manufacturing and processing of goods could be performed outside the large manufacturing plant and near to the sources of materials or the market. "Bigger no longer automatically means better: flexibility of the power unit, closer adaptation of means to ends, nicer timing of operation, are the new marks of efficient industry" (p. 226). Also see Devine, "From Shafts to Wires," p. 364; and Warren D. Devine, Jr., "Technological Change and Electrification in the Printing Industry, 1880–1930," final draft of unpublished paper, Institute for Energy Analysis, Oak Ridge Associated Universities, April 1984; and Milton F. Searl, "Electricity and Economic Growth in the United States, 1902–1982 – An Aggregative Analysis," draft paper presented at the EPRI Workshop on Electricity Use, Productive Efficiency, and Economic Growth, December 8, 1983.

18. After observing a smaller, but successful experiment performed in Los Angeles, Chicago's Commonwealth Edison began lending 10,000 electric irons to its 40,000 residence customers in 1908. "Renting 10,000 Irons in Chicago," *Electrical World* 52 (December 26, 1908): 1401.

19. Lighting use accounted for the balance in each year. William Kennedy, *The Objective Rate Plan* (New York: Columbia University Press, 1937), p. 2. Overall, utilities obtained more than 15% of their revenues from domestic appliance usage in residential buildings. See Moody's Investors Service, *Public Utilities Guide* (New York: Moody's Investors Service, 1933), p. a108.

20. Moody's Investors Service, *Public Utilities Guide* (New York: Moody's Investors Service, 1941), p. a18.

21. See Susan Strasser, *Never Done: A History of American Housework*, Chapter 4, "At the Flick of a Switch" (New York: Pantheon, 1982), pp. 67–84.

22. See the company's publication, *Harvests and Highlights* (Chicago: Middle West Utilities Company, 1930).

23. "Rural Electrification," *Electrical World* 131 (May 21, 1949): 506.

24. Marquis Childs, *The Farmer Takes a Hand: The Electric Power Revolution in Rural America* (Garden City, NY: Doubleday, 1952), p. 13. Electricity also had a profound impact on farms in other countries. For an early view of the use of electricity on farms in England, see Arthur H. Allen, *Electricity in Agriculture* (London: Pitman, 1922).

25. For a discussion of the general ideology of engineering at the beginning of the twentieth century, see Edwin T. Layton, Jr., *The Revolt of the Engineers: Social Responsibility and the American Engineering Profession* (Cleveland: Press of the Case Western Reserve University, 1971). Also see Peter Meiksins, "The 'Revolt of the Engineers' Reconsidered," *Technology and Culture* 29 (1988): 219–46.
26. Samuel Insull, "Centralization of Energy Supply," in William E. Keily, ed., *Central Station Electricity Service: Its Commercial Development and Economic Significance as Set Forth in the Public Addresses (1897–1914) of Samuel Insull* (Chicago: Private printing, 1915), p. 475.
27. Edwin Vennard, *Management of the Electric Energy Business* (New York: McGraw-Hill, 1979). Vennard was more than the normal utility manager. As vice-president and managing director of the Edison Electric Institute from 1956 to 1969, he directed campaigns to promote electricity use to unprecedented levels. A biography of Vennard is presented in the *National Cyclopedia of American Biography* (1972), pp. 542–3.
28. Government investigations of the utility industry in the late 1920s revealed that the NELA promoted more than just electricity usage and a good public image for utilities. The organization also influenced political sentiment by criticizing public power proposals and public policy that would affect utilities. Worst of all, utilities passed on the costs for NELA's propaganda efforts to customers. Due to the bad utility image that resulted from these revelations, NELA disbanded in 1933, to be replaced by the Edison Electric Institute, whose founders declared that the new industry trade group would "divest itself of any semblance of propaganda activities." Thomas K. McCraw, *TVA and the Power Fight: 1933–1939* (Philadelphia: Lippincott, 1971), p. 23. A brief proindustry history of the Edison Electric Institute can be found in "Edison Electric Institute," *Electrical World* 131 (May 21, 1949): 565–6.
29. Recently, economists and social scientists have suggested the need to consider no-growth or "steady-state" economic systems. See, for example, Herman E. Daly, ed., *Toward a Steady-State Economy* (San Francisco: W. H. Freeman, 1973); and Herman E. Daly, *Steady State Economics* (San Francisco: W. H. Freeman, 1977).
30. President's Materials Policy Commission (The Paley Commission), *Resources for Freedom. Vol. 1: Foundations for Growth and Security* (Washington, DC: U.S. Government Printing Office, 1952), p. 3.
31. Much of this discussion is taken from Ellen Maher, "The Dynamics of Growth in the Electric Power Industry," in Kenneth Sayre, ed., *Values in the Electric Power Industry* (Notre Dame, IN: University of Notre Dame Press, 1977), pp. 149–216, which itself summarizes the work of several sociologists of American culture.
32. Cited in Edwin Vennard, *The Electric Power Business*, 2nd Ed. (New York: McGraw-Hill, 1970), p. 14.
33. Cited in Thomas K. McCraw, "Triumph and Irony – The TVA," *Proceedings of the IEEE* 64 (1976): 1375.
34. See McCraw, *TVA and the Power Fight*.
35. Tennessee Valley Authority, *TVA 1982 Power Power Program Summary*, Vol. 1 (Knoxville, TN: Tennessee Valley Authority, 1983), p. 39.
36. David E. Lilienthal, *TVA: Democracy on the March* (New York: Harper, 1944), p. 17.
37. Despite suffering serious setbacks in the 1970s and 1980s as a consequence of rising costs, cancelled and shut down nuclear plants, and environmental concerns,

TVA employees are still proud of their past accomplishments. In 1983, the Authority published a lavishly illustrated book celebrating fifty years of good deeds in the power business: *OEDC Proud Fifty: The Tennessee Valley Authority Office of Engineering Design and Construction Fiftieth Anniversary Commemorative Booklet* (Knoxville, TN: Tennessee Valley Authority, 1983).
38. McCraw, *TVA and the Power Fight*, pp. 74–7.

Chapter 3

1. William R. Hughes, "Scale Frontiers in Electric Power," in William M. Capron, ed., *Technological Change in Regulated Industries* (Washington, DC: Brookings Institution, 1971), p. 48.
2. Bruce A. Smith, *Technological Innovation in Electric Power Generation, 1950–1970* (East Lansing, MI: Michigan State University, 1977), p. 12. Also see Ryutaro Komiya, "Technological Progress and the Production Function in the United States Steam Power Industry," *Review of Economics and Statistics* 44 (1962): 156–66; and Yoram Barzel, "Productivity in the Electric Power Industry," *Review of Economics and Statistics* 45 (1963): 395–407.
3. Ralph G. M. Sultan, *Pricing in the Electrical Oligopoly*, Vol. 1 (Cambridge, MA: Harvard University Press, 1974), p. 11.
4. "The Electric Utility Companies," *Electrical World* 131 (May 21, 1949): 10.
5. G. B. Warren, "Progress in Design and Performance of Modern Large Steam Turbines for Generator Drive, Part I," *General Electric Review* 43 (September 1940): 374.
6. Sultan, *Pricing in the Electrical Oligopoly*, Vol. 1, pp. 8–11.
7. See *ibid.* for more on the market and competitors in the manufacturing industry.
8. Interview with John F. Traexler, manager of Engineering, Steam Turbine Generating Division, Westinghouse Electric Corporation, April 25, 1986.
9. G. B. Warren, "Recent and Possible Future Developments Affecting the Economics of Large Steam Turbine Practice in the United States, Part I," *General Electric Review* 33 (August 1930): 435.
10. "Power Costs and High Pressure," *Electrical World* 93 (April 27, 1929): 817–18. See also St. Loeffler, "To Reduce Power Costs," *Electrical World* 93 (April 27, 1929): 829–30.
11. "Higher Temperatures Must Await Developments in Metallurgy," *Electrical World* 81 (June 23, 1923): 1459.
12. C. F. Hirshfield, "Rehabilitation of Steam Power Plants," *Electrical World* 94 (July 6, 1929): 13–16.
13. Warren, "Recent and Possible Future Developments, Part I," p. 435.
14. Ernest L. Robinson, "The Steam Turbine in the United States. III – Developments by the General Electric Company," *Mechanical Engineering* 59 (1937): 144.
15. Alvin M. Weinberg, "Science and Trans-Science" *Minerva* 10 (1972): 211.
16. Emil E. Keller and Francis Hodgkinson, "The Steam Turbine in the United States. I – Developments by the Westinghouse Machine Company," *Mechanical Engineering* 58 (1936): 687.
17. In the decade from 1910 to 1919, utilities installed a total of thirty-five fossil-fueled units. From 1920 to 1929, they installed 241, exclusive of thirteen combined-

cycle steam plants. Data from U.S. Department of Energy, Energy Information Administration, Generating Unit Reference File, February 24, 1987.

18. H. M. Hobart, "Rationalization and the Engineer," *General Electric Review* 33 (May 1930): 272.

19. *Ibid.*, p. 271.

20. Sultan, *Pricing in the Electrical Oligopoly*, Vol. 1, pp. 11–12.

21. A 1% increment in a sphere's volume increases its surface area by about 0.6%. If surface area is proportional to cost, then the rule holds. Robert J. Gordon, "The Productivity Slowdown in the Steam-Electric Generating Industry," unpublished paper, Northwestern University and National Bureau of Economic Research, October 1982, revised February 1983, p. 12. Also see F. T. Moore, "Economies of Scale: Some Statistical Evidence," *Quarterly Journal of Economies* 73 (1959): 232–45.

22. "60,000-KW Turbo-Generator for Lakeside," *Electrical World* 93 (February 23, 1929): 381–2.

23. G. B. Warren, "Recent and Possible Future Developments Affecting the Economics of Large Steam Turbine Practice in the United States, Part I," *General Electric Review* 33 (August 1930): 439.

24. Besides the scale economies achieved by increasing the size of machinery, there are scale economies associated with increasing sales. As sales of electricity increase with production coming from an existing unit or plant, the cost per unit decreases because no extra investment of capital is expended. In promoting sales to achieve this goal, utilities often found that they had to build new plants to handle the new load, which would start the cycle over again. See Alfred E. Kahn, *The Economics of Regulation: Principles and Institutions*, Vol. 1 (New York: Wiley, 1970), pp. 124–30. For economies of a different sort – economies of scale for a firm, not just a plant – see Paul Rodgers and Gordon L. Pozza, "Christensen and Greene's 'Economies of Scale in U.S. Electric Power Generation,' " National Association of Regulatory Utility Commissioners, Technical Economic Literature Review Series Paper No. 1 (July 27, 1979); and Kenneth Jameson, "Economies of Scale in the Electric Power Industry," in Kenneth Sayre, ed., *Values in the Electric Power Industry* (Notre Dame, IN: University of Notre Dame, 1977), pp. 115–48.

25. Ernest L. Robinson, "Turbine Design Trends," *Electrical World* 94 (November 16, 1929): 972. A picture of a 20-MW horizontal turbine generator can be found facing page 338 in Samuel Insull, "Supplying the Energy Requirements of the Community," in William E. Keily, ed., *Central Station Electricity Service: Its Commercial Development and Economic Significance as Set Forth in the Public Addresses (1897–1914) of Samuel Insull* (Chicago: private printing, 1915).

26. Warren, "Progress in Design and Performance, Part I," pp. 374–7.

27. Charles Edison, "New Engineering in the Navy," *Scientific American* 162 (March 1940): 138; cited in Warren, "Progress in Design and Performance, Part I," p. 376.

28. M. A. Savage, "Economic Developments in Turbine Generators in the United States," *General Electric Review* 33 (August 1930): 444.

29. The unit was designed for the State Line Generating Company of Hammond, Indiana. See R. C. Spencer, "Evolution in the Design of Large Steam Turbine-Generators," General Electric Large Steam Turbine Seminar Paper 83T1 (1983), pp. 2–3.

30. Savage, "Economic Developments," p. 445.
31. "Features of the 160,000-kW Hell Gate Turbine," *Electrical World* 93 (March 30, 1929): 625.
32. "Economic and Service Elements in Large Turbo-Generators," *Electrical World* 93 (March 30, 1929): 622.
33. Babcock and Wilcox, *The Babcock and Wilcox Story: 100 Years of Service to Industry, 1867–1967* (Barberton, OH: Babcock and Wilcox, 1967), p. 10.
34. "Growth Marks the Years 1907–57," *Power* (June 1981); in *A Century of Power Progress: A Review of the Major Technological Changes, Discoveries, and Advances that have Swept the Power World Since 1882* (New York: McGraw-Hill, 1982), p. 12.
35. Babcock and Wilcox, *The Babcock and Wilcox Story*, p. 18. More details on boilers will follow in this chapter.
36. U.S. Federal Power Commission, *National Power Survey* (Washington, DC: U.S. Government Printing Office, 1964), p. 64.
37. Cost per kilowatt data from L. W. W. Morrow, "Reduced Power Costs?" *Electrical World* 94 (September 28, 1929): 619. Average unit size data from U.S. Department of Energy, Generating Unit Reference File, February 24, 1987. Consumer price index data found in *Handbook of Basic Economic Statistics*, Vol. 40 (Washington, DC: Economic Statistics Bureau, 1986), pp. 97–101.
38. Neil, "Entering the Seventh Decade," p. 328. See also chapter entitled "White Coal" in Thomas P. Hughes, *Networks of Power: Electrification in Western Society, 1880–1930* (Baltimore: Johns Hopkins University Press, 1983).
39. P. H. Chase, W. W. Lewis, D. M. Simmons, and H. R. Woodrow, "Power Transmission and Distribution in the United States of America, Part I: Transmission," *General Electric Review* 35 (August 1932): 431.
40. According to the useful approximation, capability increases as the square of the voltage. "Electric Power Systems," *McGraw-Hill Encyclopedia of Science and Technology*, 6th Ed., Vol. 6 (New York: McGraw-Hill, 1987), p. 42.
41. A discussion of transmission problems is included in Taylor Moore, "Exploring the Limits of Power Transfer," *EPRI Journal* 13 (January/February 1988): 22–31.
42. Chase, "Power Transmission," p. 432.
43. For an excellent analysis of the crucial role played by transmission in the creation of efficient power systems, see Paul L. Joskow and Richard Schmalensee, *Markets for Power: An Analysis of Electric Utility Deregulation* (Cambridge, MA: The MIT Press, 1983), pp. 62–6.
44. Other discussions of the value of interconnection systems include Federal Power Commission, *The 1970 National Power Survey*, Part I, Chapters 13 and 17, and Part IV, Section 2 on "The Transmission of Electric Power" (Washington, DC: U.S. Government Printing Office, 1971),
45. Because the heat content of coal varies with the type of coal, the amount of impurities, and whether or not it is washed, coal rates are not completely comparable unless conversion factors are employed.
46. Babcock and Wilcox, *The Babcock and Wilcox Story*, p. 10.
47. *Ibid.*, p. 15.
48. *Ibid.*, p. 21.
49. "Boosting Powerplant Performance," *Power* (July 1981); in *A Century of Power Progress: A Review of the Major Technological Changes, Discoveries, and*

Advances that have Swept the Power World Since 1882 (New York: McGraw-Hill, 1982), p. 14.

50. Spencer, "Evolution in the Design of Large Steam Turbine-Generators," p. 2.

51. *Ibid.*

52. "The Electric Power Industry in the United States," *Electrical World* 93 (January 5, 1929): 37.

53. Charles E. Neil, "Entering the Seventh Decade of Electric Power," *Edison Electric Institute Bulletin* 10 (September 1942): 326; and Edison Electric Institute, *Edison Electric Institute Pocketbook of Electric Utility Industry Statistics*, 30th Ed. (Washington, DC: Edison Electric Institute, 1984), p. 21.

54. Electricity cost data from *ibid.*, p. 33. Calculations on cost of living and real cost of electricity based on consumer price index data found in *Handbook of Basic Economic Statistics*, pp. 97–101.

Chapter 4.

1. "Energy Marketing," *Electrical World* 181 (June 1, 1974): 368–74.

2. Edison Electric Institute, *Statistical Bulletin, Year 1949*, No. 17, Publication No. 50–2 (New York: Edison Electric Institute, 1950).

3. Frank R. Innes, "Today's Thinking on Tomorrow's Load Building," *Electrical World* 122 (November 11, 1944): 87–9; "Notes on a Power Shortage," *Electrical World* 128 (November 22, 1947): 61; and "Shortages of Reserve Capacity Tax Systems' Capabilities," *Electrical World* 128 (November 22, 1947): 74.

4. "The Electric Utility Companies," *Electrical World* 131 (May 21, 1949): 12.

5. *Ibid.*

6. "We Chose Ships," *Electrical World* 128 (November 22, 1947): 76.

7. "The Electric Utility Companies," p. 12.

8. U.S. Department of Energy, Generating Unit Reference File, February 24, 1987.

9. Virginia Electric and Power Company, for example, was divested from its holding company, Engineers Public Service Corporation (created by Stone and Webster, Inc., and four investment firms), in 1947. Erwin H. Will, *The Past – Interesting, The Present – Intriguing, The Future – Bright: A Story of Virginia Electric and Power Company* (New York: Newcomen Society of North America, 1965), pp. 15–17. (Mr. Will was chairman of the board of VEPCo.) Madison Gas and Electric likewise became a separate entity in 1946. Interview with Frederick D. Mackie, former president and general manager, MG&E, August 21, 1984.

10. R. A. Gallagher, "Why We Push the 'Go All-Electric' Program," *Electrical World* 129 (March 27, 1948): 88–9. (Mr. Gallagher was president of the Public Service Company of Indiana.)

11. "Sales Promotion," *Electrical World* 131 (May 21, 1949): 504–5.

12. Edwin Vennard, "We Must Sell Aggressively Today," *Electrical World* 131 (June 4, 1949): 80. Also see "Selective Selling: Financing Depends on Earnings – Earnings Depend on Sales," *Electrical World* 131 (March 26, 1949): 79–80.

13. "Sales Promotion," p. 505.

14. President's Materials Policy Commission (The Paley Commission), *Resources for Freedom. Vol. 1: Foundations for Growth and Security* (Washington, DC: U.S. Government Printing Office, 1952), p. 1.

15. B. L. England, "Electric Utilities Must Sell," *Electrical World* 138 (December 1, 1952): 94–5.

16. George F. Hessler, "Intensified Selling Will Pave the Road Ahead," *Electrical World* 139 (January 12, 1953): 71.

17. *John Fritz Medal Biography of Philip Sporn, Medalist for 1956* (New York: American Society of Mechanical Engineers, 1956), p. 4.

18. "Utilization of Electricity," *Electrical World* 131 (May 21, 1949): 503.

19. Moody's Investors Service, *Moody's Public Utility Manual* (New York: Moody's Investors Service, 1972), p. a20.

20. Philip Sporn, "All-Electric Home Seen At Hand," *Electrical World* 143 (April 25, 1955): 16–18.

21. "Load Research Essential Today for Tomorrow's Appliance Sales," *Electrical World* 139 (May 18, 1953): 141.

22. See, for example, "Characteristics of the All-Electric Home," *Electrical World* 145 (March 19, 1956): 123–46.

23. "Enlivening and Humanizing Electrical Advertising," *Edison Electric Institute Bulletin* 2 (September 1934): 303–6.

24. "The Man Behind the Symbol," *Reddy News* 42 (No. 948, January/February 1984), no page. *Reddy News* is a publication of Reddy Kilowatt's trademark owner, Reddy Communications, Inc.

25. Reddy Kilowatt, Inc., *The Reddy Kilowatt Story* (New York: Reddy Kilowatt, 1968), pp. 5–6.

26. Interview with Gary R. Grable, director of Advertising, Commonwealth Edison, August 20, 1984.

27. "GE to Offer Utilities Joint Market Program," *Electrical World* 144 (September 19, 1955): 112. Promotional activity since World War II and before the LBE program was on a piecemeal basis, with individual utilities and manufacturers working independently. See "Sales Promotion," *Electrical World* 131 (May 21, 1949): 504–5.

28. " 'Medallion Home' Spurs LBE Plans," *Electrical World* 148 (October 28, 1957): 47.

29. See, for example, George T. Mazuzan and J. Samuel Walker, "Developing Nuclear Power in an Age of Energy Abundance, 1946–1962," *Materials and Society* 7 (1983): 307–19. Utopian speculation about nuclear power, coming from "dreamers" and government planners, is discussed in Stephen L. Del Sesto, "Wasn't the Future of Nuclear Energy Wonderful?" in Joseph J. Corn, ed., *Imagining Tomorrow: History, Technology, and the American Future* (Cambridge, MA: The MIT Press, 1986), pp. 58–76.

30. Harlee Branch, Jr., "Sell or Die!" *Electrical World* 141 (January 18, 1954): 25. President of Georgia Power Company, Mr. Branch also noted that when "the instinct and inspiration for selling disappear, particularly when they disappear from a power company, that company is liquidating itself." "Sell – and Sell – and Sell," editorial in *Electrical World* 139 (October 6, 1952): 82. Also see the editorial, " 'Hard Sell' Has Arrived," *Electrical World* 143 (January 31, 1955): 5.

31. "How to Get Ahead in the Utilities Industry," *Business Week*, No. 1398 (June 16, 1956): 68.

32. *Ibid.*

33. Philip Sporn, "Inventing Our Future," speech at AEP Management Meeting, Wheeling, WV, November 10–12, 1964, copy provided by AEP, p. 9.

34. The basic pricing mechanism for promoting electricity use was based on Hopkinson's system of demand and energy charges. For large industrial customers who could afford the cost of the equipment, special meters measuring both the peak demand and kilowatt-hour usage accurately reflected costs. A sliding scale of declining charges applied as usage increased to spur load growth. For residential customers, however, utilities in the 1920s devised another rate schedule that did not require special devices. This was the declining block rate structure (DBR), a schedule in which the first block (or increment) of power is costly, reflecting an estimated portion of the costs incurred to serve the customer. (Demand was estimated by metering a representative sample of residential users.) Utilities usually charged for a minimum amount of usage, whether the customer used electricity or not, to pay for the available production and distribution capacity and for the costs associated with meter reading, bill collecting, and other fixed-cost services. Meanwhile, larger blocks of usage became less expensive per kilowatt-hour, as they represented primarily just the variable costs of production (i.e., the fuel). National Economic Research Associates, Inc., "An Overview of Regulated Ratemaking in the United States," report prepared for Electric Power Research Institute, February 1977, pp. 7–10. Because it promoted consumption and was simple, understandable to consumers, and representative of costs, the DBR became by the 1930s the standard rate structure used by utilities for residential service. See "Electricity Rates Explained," *Electrical World* 105 (March 30, 1935): 722–5; and Richard E. Morgan, *The Rate Watcher's Guide* (Washington, DC: Environmental Action Foundation, 1980), pp. 12–14. Virginia Electric and Power Company offered a residential rate from 1954 through 1960 that priced the first 50 kWh of electricity at 4.5 cents each, the next 50 kWh at 3.7 cents, the next 100 kWh at 2.7 cents, and any amount over 200 kWh at 1.8 cents. Through 1970, the rate structure remained almost the same, with some reductions in prices for increments after the first 50 kWh. In 1962, the company also offered reduced rates in excess of 100 kWh for homeowners having electric water heaters. Virginia Electric and Power Company, "History of Electric Rate Changes, March 1954 to February 1, 1972," unpublished VEPCo document given to the author by a VEPCo manager. Another VEPCo document noted that during the years between 1955 and 1969, "the company pursued an active policy of selectively promoting the use of electrical energy which resulted in a substantial increase in winter heating electricity consumption. . . . By 1960 the company's marketing efforts were being directed toward the all-electric living concept." "Historical Development of the Rate Structure of Virginia Electric and Power Company," unpublished VEPCo document dated 1976.

35. Paul H. Coortner and George O. G. Lof, *Water Demand for Steam Electric Generation: An Economic Projection Model* (Washington, DC: Resources for the Future, 1965 [distributed by Johns Hopkins University Press]), p. 4.

36. Data from U.S. Department of Energy, Energy Information Administration, Generating Unit Reference File, February 24, 1987.

37. Data from Edison Electric Institute, *Edison Electric Institute Pocketbook of Electric Utility Industry Statistics*, 27th Ed. (Washington, DC: Edison Electric Institute, 1981), p. 17; "1957 Annual Statistical Report," *Electrical World* 143 (January 28, 1957): 133; and "1966 Annual Statistical Report," *Electrical World* 165 (February 21, 1966): 89.

38. Virginia Electric and Power Company, *Annual Report, 1968*, pp. 22–3.

39. "Nuclear Intentions Rise as New Capacity Plans Rocket to 97.5 Million KW by 1972," *Electrical World* 165 (February 21, 1966): 84–5.

40. See G. B. Warren, "Recent and Possible Future Developments Affecting the

Economics of Large Steam Turbine Practice in the United States, Part I," *General Electric Review* 33 (August 1930): 439; and Charles A. Powel, *Principles of Electric Utility Engineering* (Cambridge, MA: The MIT Press, 1955), p. 62.

41. AEP, "1969 Edison Award Entry," no page. Some utilities exceeded the rule slightly, as was the case with Detroit Edison in 1955. As its system approached 2,500 MW, the firm ordered a 260-MW unit. L. K. Kirchmayer, A. G. Mellor, J. F. O'Mara, and J. R. Strevenson, "An Investigation of the Economic Size of Steam-Electric Generating Units," *AIEE Transactions* 74 Part 3 (1955): 611. New York's Consolidated Edison overstepped even the 10% rule in 1961 by ordering a unit that yielded 1,000 MW at a time when its system capacity was only 6,000 MW. By the time the unit would be installed, however, the base capacity (exclusive of the new unit) would have grown to 6,500 MW. Thus, the new unit would constitute 15% of the system's capacity. "Con Edison Will Buy Generator to Produce One Million Kilowatts," *Wall Street Journal* (September 28, 1961), p. 30. As will be noted later, the new unit did not operate well, causing severe problems for the utility and its customers. In the summer of 1970, a second breakdown of the unit, which then constituted 13% of the utility's capacity, occasioned *The New York Times* to criticize the company's management for relying on "one mammoth machine of questionable reliability, a machine which takes days to tear down merely to locate the source of trouble, a machine which make take months to repair." Editorial, "Out of Order," *The New York Times* (July 25, 1970), p. 22.

42. Interview with Walter L. Marshall, manager, Turbine Technology Laboratory, General Electric Company, Schenectady, NY, May 22, 1984; and interview with James M. Lydon, executive vice president, Boston Edison Company, May 23, 1985.

43. Telephone interview with Charles Elton, former manager of Large Steam Turbine Business, General Electric, November 19, 1982.

44. "New Capacity Plans Hit Record 64.5 Million KW," *Electrical World* 161 (February 24, 1964): 81.

45. Leonard M. Olmsted, "Utilities Build Big Units to Gain Economies of Scale," *Electrical World* 170 (October 21, 1968): 83.

46. Edison Electric Institute, *Edison Electric Institute Pocketbook of Electric Utility Industry Statistics*, 27th Ed. (Washington, DC: Edison Electric Institute, 1981), pp. 22–3. From $6.94 in 1957, the cost per ton of coal equivalent dropped to $6.22 in 1967 – a decrease of about 10%. *Ibid.*, p. 22. The average cost decline, however, includes nuclear fuel. The cost of individual fossil fuels as burned at steam electric power plants between 1957 and 1967 showed current and constant price patterns as follows: coal – down 8.4% current, 22.8% constant; residual oil – down 27.5% current, 38.9% constant; and gas – up 26.7% current, 6.8% constant. Fuel data from National Coal Association, *Steam-Electric Plant Factors* (Washington, DC: National Coal Association, issues from 1959–64, 1966, 1967, 1969, and 1970). Calculations of constant costs determined by adjusting the nominal prices by the consumer price index as published in *Handbook of Basic Economic Statistics*, Vol. 40 (Washington, DC: Economic Statistics Bureau, 1986): 97–101.

47. Richard H. K. Vietor, *Energy Policy in America Since 1945* (Cambridge: Cambridge University Press, 1984), pp. 91–115.

48. Sam H. Schurr and Bruce C. Netschert, *Energy in the American Economy, 1850–1975* (Baltimore: Johns Hopkins University Press, 1960), pp. 77–81.

49. The index is a composite of more than fifty material and labor items for steam power plants in different geographical regions of the United States. The figures used are for the North Atlantic region and were compiled by Whitman, Requart, and Associates, Baltimore. 1949 is the base year (= 100). From Leonard S. Hyman, *America's Electric Utilities: Past, Present, and Future* (Arlington, VA: Public Utility Reports, 1983), pp. 90 and 110.

50. Edison Electric Institute, *Pocketbook*, p. 22. The TVA in 1970, for example, had become worried about coal deliveries, contributing to its desire to move toward nuclear power as a way to reduce costs.

51. Interview with Dwight Patterson, assistant to the TVA manager of Engineering Design, January 18, 1984.

52. This is an expression commonly heard at Commonweath Edison Company. Interview with Robert L. Bolger, assistant vice president, Commonwealth Edison Company, August 20, 1984.

53. Philip Sporn, "Prospects for Cost Reductions," speech at World Power Conference, Lausanne, Switzerland, September 17, 1964, in Philip Sporn, *Vistas in Electric Power* Vol. 1 (London: Pergamon, 1968), pp. 194–5.

54. Ray Schuster, "Siting Sorrows," *Power Engineering* (December 1970): 29.

55. *Ibid.*

56. Interview with George Palo, former TVA manager of Engineering Design and Construction, January 17, 1984. Also interview with Herman Koenig, former chairman of Electrical Engineering at Michigan State University, December 14, 1983.

57. James Cook, "Nuclear Follies," *Forbes* 135 (February 11, 1985): 84.

58. For a description of the rapid growth of nuclear units' capacities, see Irvin C. Bupp and Jean-Claude Derian, *Light Water: How the Nuclear Dream Dissolved* (New York: Basic Books, 1978).

59. L. K. Kirchmayer, A. G. Mellor, J. F. O'Mara, and J. R. Stevenson, "An Investigation of the Economic Size of Steam-Electric Generating Units," *AIEE Transactions* 74 Part 3 (1955): 600–9.

60. See the discussion of Kirchmayer, *et. al.*'s paper, which follows in *ibid.*, pp. 609–14. Also see "Power for Tomorrow," *General Electric Review* 56 (September 1953): 10. Economies of scale in the 1950s were also noted in the President's Materials Policy Commission (The Paley Commission), *Resources for Freedom. Vol. 3: The Outlook for Energy Sources* (Washington, DC: U.S. Government Printing Office, 1952), p. 35.

61. One study that focused on plant size increases from 1929–55, for example, showed that for every 10% enlargement of size, productivity increased 1.09%. Put another way, the cost of producing 1 kWh dropped 1.09% for each 10% increase in size. See Yoram Barzel, "Productivity in the Electric Power Industry," *Review of Economics and Statistics* 45 (1963): 402. Other studies suggested that a 10% increase in size led to cost increases for the largest units of 8% in the 1950s and 9% in the 1960s. William R. Hughes, "Scale Frontiers in Electric Power," in William M. Capron, ed., *Technological Change in Regulated Industries* (Washington, DC: Brookings Institution, 1971), p. 48. Also see Ryutaro Komiya, "Technological Progress and the Production Function in the United States Steam Power Industry," *Review of Economics and Statistics* 44 (1962): 156–66. For more on studies of economies of scale, see T. W. Schroeder and G. P. Wilson, "Economic Selection of Generating Capacity Additions," *AIEE Transactions, Part III, Power Apparatus and Systems* 77 (1958): 1133–45; Marc Nerlove, "Returns to Scale in Electricity Supply," in Carl F. Christ, *et. al.*, *Measurement in Economics* (Stanford, CA: Stanford University Press, 1963), pp. 167–98; Clifford F. Pratten, *Economies of Scale in*

Manufacturing Industry (Cambridge: Cambridge University Press, 1975), pp. 197–207; and Malcolm Galatin, *Economies of Scale and Technological Change in Thermal Power Generation* (Amsterdam: North-Holland, 1968.) More recent studies consist of Wesley D. Seitz, "Productive Efficiency in the Steam Electric Generating Industry," *Journal of Political Economy* 79 (1971): 878–86; Thomas G. Cowing, "Technical Change and Scale Economies in an Engineering Production Function: The Case of Steam Electric Power," *Journal of Industrial Economics* 23 (December 1974): 134–52; Wilfred H. Comtois, "Economy of Scale in Power Plants," *Power Engineering* 81 (August 1977): 51–3; Kenneth Jameson, "Economies of Scale in the Electric Power Industry," in Kenneth Sayre, ed., *Values in the Electric Power Industry* (Notre Dame, IN: University of Notre Dame Press, 1977): 116–48; A. J. Abdulkarim and N. J. D. Lucas, "Economies of Scale in Electricity Production in the United Kingdom," *Energy Research* 1 (1977): 223–31; and Rajat K. Deb and James J. Mulvaney, "Economic and Engineering Factors Affecting Generating Unit Size," *IEEE Transactions on Power Apparatus and Systems* PAS-101 (October 1982): 3907–18. Other useful information concerning scale economies came from the Tennessee Valley Authority, which in the 1950s and 1960s built some large pioneering plants. Because the TVA constructed its units without help from outside architect-engineering firms, fewer outside variables affected its experience. See G. O. Wessenauer, W. E. Dean, Jr., and J. E. Gilleland, "Some Problems on a Large Power System with Large Generating Units," paper presented at the Sixth World Power Conference, Melbourne, Australia, October 20–27, 1962, pp. 4–5; and Roy H. Dunham, "Growth of Steam Power of TVA System," *Power Engineering* 74 (July 1970): 40.

62. Utilities also hoped to modernize and automate their plants. "Manufacturer Shipments Jumped 11.5% to $19.6 Billion in 1956," *Electrical World* 147 (January 28, 1957): 142–3.

63. Much of this discussion draws from evidence gathered during collusion trials in the early 1960s and is described in detail in Ralph G. M. Sultan, *Pricing in the Electrical Oligopoly*, Vols. 1 and 2 (Cambridge, MA: Harvard University Press, 1974 and 1975). This information comes from Vol. 1, pp. 167–9.

64. Sultan, *Pricing in the Electrical Oligopoly*, Vol. 2, pp. 178–80.

65. *Ibid.*, pp. 174 and 223.

66. For a sense of the overwhelming task taken on by Westinghouse in order to compete with GE, see the remarks made by D. W. R. Morgan, general manager of Westinghouse Steam Division, 1948–53, and John R. Carlson, division engineering manager at Westinghouse's Lester plant, in the case of *Ohio Valley Electric v. General Electric*, in Sultan, *Pricing in the Electrical Oligopoly*, Vol. 2, p. 175.

67. U.S. Federal Power Commission, *National Power Survey*, Part 1 (Washington: DC: U.S. Government Printing Office, 1964), p. 65.

68. Interview of J. David Conrad, manager, Business Development, Power Generation, Westinghouse Electric Corporation, August 9, 1985.

69. Interview with John F. Traexler, manager of Engineering, Westinghouse Steam Turbine Generator Division, April 25, 1986.

70. Sultan, *Pricing in the Electrical Oligopoly*, Vol. 1, p. 184. This discussion of GE is based primarily on Sultan's work, pp. 183–5.

71. Interview of Eugene J. Cattabiani, executive vice president, Power Generation, Westinghouse Electric Corporation, July 29, 1985.

72. Data from National Coal Association, *Steam-Electric Plant Factors, 1959* (Washington, DC: National Coal Association, 1960), p. 25.
73. Interview with Joseph Swidler, former attorney for the TVA, former chairman of the Federal Power Commission, and former chairman of the New York State Public Service Commission, March 15, 1983.
74. Interview of J. David Conrad.
75. In general, Allis-Chalmers sold small units, with 76% of its sales in the 1948-to-1962 period in the size range below 100 MW. Sultan, *Pricing in the Electrical Oligopoly*, Vol. 2, p. 204. Meanwhile, the Tennessee Valley Authority had placed an order for a 950-MW plant before the Consolidated Edison order was announced. "Con Edison Will Buy Generator to Produce One Million Kilowatts," *Wall Street Journal* (September 28, 1961), p. 30; and "Con Ed Orders First 1,000-MW Unit," *Electrical World* 156 (October 2, 1961): 63.
76. This view was presented to the author by Joseph Pratt, a business historian at Texas A&M University, who has written an unpublished corporate history of Consolidated Edison. Telephone interview, March 15, 1984.
77. "Con Edison's 1,000 MW Unit on Line," *Electrical World* 163 (June 14, 1965): 69–70. Part of the reduced cost reflected the large discount offered by Allis-Chalmers.
78. "Conventional Steam-Electric Generating Stations," Advisory Committee Report No. 7, April 1963, Part 2, p. 58 of *National Power Survey*.
79. *Ibid.*
80. Babcock and Wilcox, *Steam: Its Generation and Use* (New York: Babcock and Wilcox, 1978), Chapter 1, pp. 6–7.
81. A. J. Surrey and J. H. Chesshire, *The World Market for Electric Power Equipment: Rationalisation and Technical Change* (Sussex: University of Sussex Science Policy Research Unit, 1972), pp. 162–4.
82. "Power for Tomorrow," p. 10.
83. R. C. Spencer, "Evolution in the Design of Large Steam Turbine-Generators," General Electric Large Steam Turbine Seminar paper 83T1, 1983, pp. 13–14.
84. *Ibid.*, p. 10.
85. Most of this information comes from an interview with Norman H. Jones, manager, General Electric Company, Generator Engineering Section, May 23, 1984.
86. "Power for Tomorrow," p. 11. Also see "First Large Generator with Liquid Cooled Conductors," *General Electric Review* 56 (March 1953): 41.
87. Interview with Norman H. Jones. Also see John S. Joyce, Wilhelm Engelke, and Dietrich Lambrecht, "Will Large Turbine-Generators of the Future Require Superconducting-Field Turbogenerators?" *Proceedings of the American Power Conference* 39 (1977): 257.
88. A. F. Armor, "Efficiency Improvement in Fossil-Fired Power Plants," American Society of Mechanical Engineers paper 83-JPGC-Pwr-5 (1983), p. 2. Similar comments are made by the same author in "Power Plant Efficiency Improvement: An EPRI Perspective," in D. E. Leaver and R. G. Brown, *Fossil Plant Heat Rate Improvements: 1981 Conference and Workshop*, publication CS-2180 (Palo Alto, CA: Electric Power Research Institute, 1981), pp. 1–4.
89. Spencer, "Evolution in the Design of Large Steam Turbine-Generators," p. 7.
90. Babcock and Wilcox, *Steam*, Chapter 21, p. 16.
91. *Ibid.* Also see Rajat K. Deb and James J. Mulvaney, "Economics and Engi-

neering Factors Affecting Generating Unit Size," *IEEE Transaction on Power Apparatus and Systems* PAS-101 (October 1982): 3907.
92. U.S. Federal Power Commission, *National Power Survey*, p. 83. Other studies in the 1960s reinforced this view. One evaluation of construction costs for coal-fired generating units and nuclear generating units suggested that while construction costs for a coal-fired plant would decrease 6.5% on jumping from 580 to 870 MW, a nuclear plant's cost would decrease 15.4%. Another step to 1,160 MW would reduce the conventionally fueled plant's cost by 2.2%, whereas a nuclear plant would cost 9.1% less. See C. Maxwell Stanley, "The Impact of Changing Economics on Electric Utilities," *Proceedings of the American Power Conference* 31 (1969): 12–13.
93. U.S. Federal Power Commission, *National Power Survey*, pp. 82–3.
94. Bupp, *Light Water*, pp. 73–4.
95. Leonard S. Hyman, *America's Electric Utilities: Past, Present, and Future* (Arlington, VA: Public Utilities Reports, 1983), p. 90. When *Electrical World* performed its fifteenth "Steam Station Cost Survey" of the twenty new stations (containing twenty-three units) put in operation during 1964, 1965, and early 1966, it found that the smallest stations – 75 to 99 MW large – cost over $160 per kilowatt. Larger stations – ranging from 300 to 999 MW – cost under $120 per kilowatt. "Investment down 7.9%; Heat-Rate Drop Sets Record," *Electrical World* 168 (October 16, 1967): 108.
96. B. A. Smith, *Technological Innovation in Electric Power Generation, 1950–1970* (East Lansing, MI: Michigan State University, 1977), p. 16.

Chapter 5

1. Arthur L. Donovan, "Engineering in an Increasingly Complex Society," in National Research Council, *Engineering in Society* (Washington, DC: National Academy Press, 1985), p. 110.
2. *Ibid*.
3. *Ibid*., p. 111.
4. Emphasis added. Charles F. Scott, cited in A. Michal McMahon, *The Making of a Profession: A Century of Electrical Engineering in America* (New York: IEEE Press, 1984), p. 108.
5. Samuel Insull, "Twenty-Five Years of Central Station Commercial Development," in William E. Keily, ed., *Central Station Electricity Service: Its Commercial Development and Economic Significance as Set Forth in the Public Addresses (1897–1914) of Samuel Insull* (Chicago: private printing, 1915), p. 151.
6. For a discussion of how electrical engineers entered management ranks in the early twentieth century, see McMahon, *The Making of a Profession*, Chapter 4, pp. 99–132. Specific reference to engineering management in electric utilities is on pp. 99–112.
7. "Utilities Boost Recruiting Goals During Demand Ease-ups," *Electrical World* 162 (October 12, 1964): 33.
8. Ellen Maher, "The Dynamics of Growth in the Electric Power Industry," in Kenneth Sayre, ed., *Values in the Electric Power Industry*, (Notre Dame, IN: Univeristy of Notre Dame Press, 1977), p. 170.
9. Edward Constant, II, *The Origins of the Turbojet Revolution* (Baltimore: Johns Hopkins University Press, 1980), pp. 8–10.
10. E. W. Norris, M. J. Lowenberg, and T. E. Penard, "Pioneer Engineering at

Weymouth Station," *Electrical World* 85 (April 18, 1925): 808–15; and Edison Electric Institute, *Edison Electric Institute Pocketbook of Electric Utility Industry Statistics*, 30th Ed. (Washington, DC: Edison Electric Institute, 1984), p. 21.

11. Norris, "Pioneer Engineering," p. 809.

12. "Higher Steam Temperatures," *Electrical World* 93 (June 22, 1929): 1272–3.

13. "Size Limits in Turbo-Generators," *Electrical World* 92 (December 29, 1928): 1283.

14. Interview with John J. Viera, vice president, Commonwealth Edison Company, August 20, 1985; and Commonwealth Edison, "A History of Electric Service in Chicagoland" (January 1983): 10. ("A History" is an internal company document.)

15. It operated seven plants in 1984. Kenneth Labich, "Premature Obituary for Nuclear Power," *Fortune* 109 (February 20, 1984): 123.

16. Interview with George Kimmons, TVA manager of Engineering Design and Construction, January 17, 1984. Also see Thomas K. McCraw, "Triumph and Irony – The TVA," *Proceedings of the IEEE* 64 (1976): 1376.

17. AEP Service Corporation, "Pioneering in Electric Power Technology," (Columbus, OH: AEP Service Corporation, 1984), p. 17.

18. "AEP to Repower Sporn Plant with PFBC," *Electrical World* 202 (June 1988): 13.

19. Ralph G. M. Sultan, *Pricing in the Electrical Oligopoly*, Vol. 2 (Cambridge, MA: Harvard University Press, 1975), p. 199.

20. Bruce A. Smith, *Technological Innovation in Electric Power Generation, 1950–1970* (East Lansing, MI: Michigan State University, 1977), p. 57. More than one manager told me that utility presidents especially wanted the biggest and best technologies in plants because they would eventually carry the chief executive's name.

21. Of course, utilities did not build the technology themselves; they used it within a system that they designed for their own specific needs. Nevertheless, manufacturers encouraged this credit-taking behavior because it helped them sell the equipment. Interview with Eugene J. Cattabiani, executive vice president, Power Generation, Westinghouse Electric Corporation, July 29, 1985.

22. Philip Sporn, *The Social Organization of Electric Power in Modern Societies* (Cambridge, MA: The MIT Press, 1971), pp. 94–5.

23. Forrest McDonald, *Insull* (Chicago: University of Chicago Press, 1962), pp. 98–101.

24. Interview with George B. Cox, senior vice president and group executive, Turbine Business Group, General Electric, May 21, 1984.

25. Interviews with Aubrey J. "Red" Wagner, former chairman of the board of TVA, January 18, 1984; and George B. Cox. One source suggests that the TVA went to the Parsons Company for a 500-MW turbine generator for the Colbert steam plant in 1959 because its managers felt that American companies were setting prices too high and that it could force lower prices by buying abroad. The TVA action was the initial salvo in what became the famous Congressional hearings and federal anticollusion trials of 1960. See Clarence C. Walton and Frederick W. Cleveland, Jr., *Corporations on Trial: The Electric Cases* (Belmont, CA: Wadsworth, 1964), p. 29; and Sultan, *Pricing in the Electrical Oligopoly*, Vol. 2, pp. 84–6.

26. In 1959, the British installed its first unit in the range between 181 and 360 MW. An American company had already built a 500-MW unit. Not until 1966 did the CEGB add its first 500-MW unit. British data from Central Electricity Generating Board, Generation Development and Construction Division, *Advances in Power Station Construction* (Oxford: Pergamon Press, 1986), pp. 4–7. In France, a similar program of stepped standardization occurred, with construction of *only* 125-MW units from 1955 to 1961. See Leslie Hannah, *Engineers, Managers, and Politicians: The First Fifteen Years of Nationalised Electricity Supply in Britain* (Baltimore: Johns Hopkins University Press, 1982), p. 115. During the 1980s, France still took this approach with its nuclear power-plant construction program. The lack of standardization in American plants was also described to the author by Hugh G. Parris, TVA manager of Power, January 19, 1984.

27. Hughes, "Scale Frontiers," p. 45

28. Interview with John E. Dolan, vice chairman, Engineering and Construction, American Electric Power Service Corporation, July 30, 1985.

29. Interview with George B. Cox, senior vice president and group executive, Turbine Business Group, General Electric, May 21, 1984.

30. Sporn explained this philosophy in a paper, "Perils and Profits of Pioneering," presented to the American Institute of Electrical Engineers, Oak Ridge, TN, April 22, 1955, found in Philip Sporn, *Vistas in Electric Power*, Vol. 1 (Oxford: Pergamon Press, 1968), pp. 136–44. The approach had other benefits as well. For example, by ordering a series of units, the company could negotiate better purchase agreements with the manufacturers. Meanwhile, it could reduce its financial risk to some extent by buying backup spare parts for the entire series of units over a period of several years. Instead of purchasing backup parts for a single machine and inventorying them, the company would purchase only a percentage of the parts needed for all the machines. The inventory would therefore be used more efficiently, permitting it to "take the shakedown risks of developing new technology and [building] a higher level of excellence in the system." Interview with John E. Dolan.

31. Information provided by AEP Service Corporation, Columbus, OH. Winning an award given by an electric manufacturer can be considered incestuous and not terribly meaningful outside the power industry. But to managers of the day, the award held special meaning.

32. AEP Service Corporation, "Pioneering in Electric Power Technology."

33. Duke Power has maintained an excellent reputation for its relatively successful completion and operation of nuclear plants. See Gordon D. Friedlander, "A Utility Earns High Marks," *Electrical World* 198 (January 1984): reprint page 1; and Ed Bean, "Duke Power Succeeds in Building Nuclear Units Without Outside Help," *Wall Street Journal* (October 17, 1984), pp. 1 and 24. Also see sidebar article, "The Best," in James Cook, "Nuclear Follies," *Forbes* 135 (February 11, 1985): 93.

34. Interview with Christopher C. Rolfe, manager, Research and Projects, Design Engineering, Duke Power Company, August 8, 1985.

35. AEP, "1969 Edison Award Entry," at the invitation of Edison Electric Institute, April 15, 1970, provided by AEP Service Corporation, Columbus, OH. Another example of risk sharing consists of utilities working with manufacturers in constructing nuclear power plants. See Walton, *Corporations on Trial*, pp. 112–13.

36. Interview with Perry G. Brittain, chairman of the board and chief executive officer, Texas Utilities, January 14, 1985.

37. Paul L. Joskow and Richard Schmalensee, *Markets for Power: An Analysis of Electric Utility Deregulation* (Cambridge, MA: The MIT Press, 1983), p. 47.

38. Harvey Averch and Leland L. Johnson, "Behavior of the Firm Under Regulatory Constraint," *American Economic Review* 52 (1962): 1052–69.

39. The importance of the Averch-Johnson effect in regulated industries has been hotly debated. It has been pointed out, for example, that the use of new capital-intensive technologies may "pad" the rate base, but it may also have a counteracting effect if it also results in lower unit costs of electricity. In a regulated industry, the company that reduces its costs would also be forced to reduce its rates to customers and give up some of the benefits of the new technology. In other words, the new technology would not benefit the regulated company as much as an unregulated company, which would receive the total impact of lower costs by making higher profits (assuming it does not reduce the price of the product). Meanwhile, due to methodological difficulties in testing the Averch-Johnson effect, its validity still has not been firmly established. For discussions of the effect, see Leland L. Johnson, "The Averch-Johnson Hypothesis after Ten Years," in William G. Shepherd and Thomas G. Gies, eds., *Regulation in Further Perspective: The Little Engine that Might* (Cambridge, MA.: Ballinger, 1974), pp. 67–78; and Sunit K. Khanna, "Economic Regulation and Technological Change: A Review of the Literature," *Public Utilities Fortnightly* 109 (January 21, 1982): 37–8. An excellent discussion of the hypothesis – explaining its merits and deficiencies – is included in Alfred E. Kahn, *The Economics of Regulation: Principles and Institutions*, Vol. 2 (New York: Wiley, 1971), pp. 49–59 and 106–109. A more recent summary of the points of debate is included in Claire Holton Hammon, "An Overview of Electric Utility Regulation," in John C. Moorhouse, ed., *Electric Power: Deregulation and the Public Interest* (San Francisco: Pacific Research Institute for Public Policy, 1986), pp. 53–4.

40. Robert B. Reich, *The Next American Frontier* (New York: Times Books, 1983). A similar view about the use of mass-production techniques as a paradigm of industrial management can be found in Michael J. Piore and Charles F. Sabel, *The Second Industrial Divide: Possibilities for Prosperity* (New York: Basic Books, 1984).

41. Utility managers did not act like their counterparts in other industries in the 1950s and 1960s, however, when those industries balked at introducing technological innovations because they already had invested heavily in older processes. This concern for short-term benefits (accruing from reduced capital outlays) may have cost American industries dearly as foreign producers employed novel technologies, such as basic oxygen furnaces and continuous casting in steel manufacturing, that were neglected in the United States. Reich, *The Next American Frontier*, p. 91. Unlike managers in these enterprises, electric utility executives did not give up their quest for incrementally better and larger-scale technology. To be sure, utility managers needed the equipment to meet increasing demand, but they also wanted advanced machinery because it fulfilled their desires as engineers.

42. F. R. Harris, "The Parsons Centenary – A Hundred Years of Steam Turbines," *Proceedings of the Institution of Mechanical Engineers* 198A (1984): 214.

43. Concerning the American Electric Power Company's construction of larger and more advanced plants, *Forbes* magazine commented in 1960 that "it has been just this single-minded concentration on the bigger and better which has brought it a money-making combination of self-generating growth and painless financing." "The Glittering Lights of American Electric," *Forbes* 86 (October 1, 1960): 19.

Chapter 6

1. Moody's Electric Utility average peaked at 119.57 (monthly figure) in April 1965, up from an mean of 57.96 during the entire year of 1958. At the height of the stock market euphoria in August 1929, the average stood at 169.80, from which point it declined to 11.41 in April 1942. Moody's Investors Service, *Public Utility Manual* (New York: Moody's Investors Service, 1985), p. a-10.
2. H. H. Landsberg and S. H. Schurr, *Energy in the United States: Sources, Uses, and Policy Issues* (New York: Random House, 1968), p. 63.
3. This figure is for the net national product increase between 1900 and 1960. H. N. Schreiber, H. G. Vatter, and H. V. Faulkner, *American Economic History* (New York: Harper and Row, 1967), p. 14.
4. Charles E. Neil, "Entering the Seventh Decade of Electric Power," *Edison Electric Institute Bulletin* 10 (September 1942): 326; and Edison Electric Institute, *Edison Electric Institute Pocketbook of Electric Utility Industry Statistics*, 30th Ed. (Washington, DC: Edison Electric Institute, 1984), p. 21.
5. A. J. Surrey and J. H. Chesshire, *The World Market for Electric Power Equipment: Rationalisation and Technical Change* (Sussex: University of Sussex Science Policy Research Unit, 1972), pp. 161–5.
6. John W. Kendrick, *Productivity Trends in the United States* (Princeton: Princeton University Press for National Bureau of Economic Research, 1961), pp. 137–9. In the periods from 1953–7, 1957–60, and 1960–6, total factor productivity in the electric and gas utility group increased at annual rates of 5.6, 4.3, and 4.7%. These rates compare to increases of 1.9, 2.3, and 2.9% for the entire private domestic business economy during the same intervals. John W. Kendrick, *Postwar Productivity Trends in the United States, 1948–1969* (New York: Columbia University Press, 1973), pp. 78–9. For other studies of productivity in the utility industry, see Randy A. Nelson and Mark E. Wohar, "Regulation, Scale Economies, and Productivity in Steam-Electric Generation," *International Economic Review* 24 (1983): 57–79; Frank M. Gollop and Dale W. Jorgenson, "U.S. Productivity Growth by Industry, 1947–1973," in J. W. Kendrick and B. N. Vaccara, eds., *Studies in Income and Wealth* (Chicago: University of Chicago Press, 1980), pp. 17–136; Frank M. Gollop and Mark J. Roberts, "The Sources of Economic Growth in the U.S. Electric Power Industry," in Thomas G. Cowing and Rodney E. Stevenson, eds., *Productivity Measurement in Regulated Industries* (New York: Academic Press, 1981), pp. 107–43; Frank M. Gollop and Mark J. Roberts, "Environmental Regulations and Productivity Growth: The Case of Fossil-Fueled Electric Power Generation," *Journal of Political Economy* 9 (1983): 654–74; and John W. Kendrick, *Interindustry Differences in Productivity Growth* (Washington, DC: American Enterprise Institute, 1983). For a discussion of productivity in the early history of the utility industry, see Jacob M. Gould, *Output and Productivity in the Electric and Gas Utilities, 1899–1942* (New York: National Bureau of Economic Research, 1946).
7. *Charlotte Observer*, June 6, 1960.
8. Industrial users saw a smaller relative price drop, but their unit cost was less than half that of residential customers.
9. Virginia Electric and Power Company, *Annual Reports*, 1961 and 1972. The share price calculation is based on the average of the annual high and low prices for 1951 ($6.96) and 1967 ($44.06) as adjusted for stock splits.
10. According to the chairman of the FPC, the report was prepared by the mem-

bers, staff, and consultants of the Commission. A draft of the report circulated among members of the FPC's Executive Advisory Committee, composed of industry representatives, and was extensively rewritten taking into account those industry comments that appealed to the Commission. The appendices of the report, in contrast, were authored by industry representatives without being substantially edited by the Commission members. Letter from Joseph C. Swidler, former FPC chairman, to author, March 18, 1983.

11. U.S. Federal Power Commission, *National Power Survey* (Washington, DC: U.S. Government Printing Office, 1964), p. 71.

12. One utility consulting group published a graph in 1967 showing unit capacity reaching 1,500 MW in 1976, 2,000 MW in 1980, and 3,000 MW in 1984. See Ebasco Services, *1967 Business and Economic Charts* (New York: Ebasco Services, 1967), p. 31.

13. U.S. Federal Power Commission, *National Power Survey,* p. 1.

14. Richard L. Gordon, *Reforming the Regulation of Electric Utilities* (Lexington, MA: D. C. Heath, 1982), pp. 133–4.

15. A view of the benign nature of electric utility regulation and the environment in which it persisted is included in Charles F. Phillips, Jr., *The Regulation of Public Utilities: Theory and Practice* (Arlington, VA: Public Utilities Reports, 1984), p. 14.

16. Cited is Howard Perry of the Department of Energy in Tom Alexander, "The Surge to Deregulate Electricity," *Fortune* 104 (July 13, 1981): 98. The chairman of the Public Service Commission of West Virginia noted in 1972 that until recently, the efficiency of utility companies "made the job of the regulatory commissions the relatively simple one of approving rate reductions." Elizabeth V. Hallanan, "The Real Crisis," in IEEE, *The Commissioner's Role in Electric Power Research*, Special Publication No. 5 of the IEEE Power Engineering Society, presented at the 1972 Summer Meeting, San Francisco, July 11, 1972, IEEE document 72 CH0722-9-PWR, p. 3.

17. Douglas D. Anderson, *Regulatory Politics and Electric Utilities: A Case Study in Political Economy* (Boston: Auburn House, 1981), p. 68.

18. *Ibid.*, p. 180.

19. See Claire Holton Hammon, "An Overview of Electric Utility Regulation," in John C. Moorhouse, ed., *Electric Power: Deregulation and the Public Interest* (San Francisco: Pacific Research Institute for Public Policy, 1986), p. 47; and Paul L. Joskow and Richard Schmalensee, *Markets for Power: An Analysis of Electric Utility Deregulation* (Cambridge, MA: The MIT Press, 1983), p. 7.

20. The notion of a contract or compact existing between the regulated utilities and society is reflected in Vincent Butler, "A Social Compact to Be Restored," *Public Utilities Fortnightly* 116 (December 26, 1985): 17–21; and Irwin M. Stelzer, "The Utilities of the 1990s," *Wall Street Journal* (January 7, 1987), p. 20.

21. These comments are not meant to suggest that utilities avoided becoming embroiled in controversy. For some utilities, especially small ones (such as Madison Gas and Electric), debate continued from the 1930s through the 1970s concerning whether the companies should be bought by municipal or other government authorities. After the holding company abuses of the 1930s, the "buy-out" option frequently received much public attention as a way to avoid the potential problems arising from stockholder ownership. Interviews with managers of Madison Gas and Electric Company, August 21, 1984; and Alfred Kahn, former chairman, New York State Public Service Commission, November 9, 1983.

Chapter 7

1. The thermal efficiencies of both fossil and nuclear plants are included in these data. Only 0.4% of all fuel used to produce electricity in 1965 (and 0% in 1947) was nuclear. Edison Electric Institute, *Edison Electric Institute Pocketbook of Electric Utility Statistics*, 30th Ed. (Washington, DC: Edison Electric Institute, 1984), pp. 21, 23, and 24.
2. William R. Hughes, "Scale Frontiers in Electric Power," in William M. Capron, ed., *Technological Change in Regulated Industries* (Washington, DC: Brookings Institution, 1971), p. 61.
3. *Ibid.*, pp. 62–3.
4. The unit's heat rate of 8,500 Btu/kWh was 500 Btu/kWh less than the best subcritical unit. Bruce A. Smith, *Technological Innovation in Electric Power Generation, 1950–1970* (East Lansing, MI: Michigan State University, 1977), p. 62. Besides fuel economy, expected advantages of converting water to steam in supercritical boilers included reduced corrosion of parts, due to less film boiling on the surfaces of tubes; less erosion of turbine blades from "vapor pitting"; and reduction in the size of plant components, especially the superheater, pipes, and primary turbine stages. K. T. Monrose, "3500 PSIG Operation Should Provide Greater Reliability," *Power Engineering* (December 1963): 37–9; cited in Kenneth J. Morgan, "Electrical Utilities," in John E. Ullmann, *The Improvement of Productivity: Myths and Realities* (New York: Praeger, 1980), p. 311.
5. R. C. Spencer, "Evolution in the Design of Large Steam Turbine-Generators," General Electric Large Steam Turbine Seminar, Paper 83T1, 1983, p. 9; and "Power for Tomorrow," *General Electric Review* 56 (September 1953): 10.
6. Edison Electric Institute, *Pocketbook*, p. 21.
7. U.S. Energy Information Administration, *Thermal-Electric Plant Construction Cost and Annual Production Expenses–1980* (Washington, DC: Department of Energy, 1983), p. 12.
8. Data for the newly installed plants come from U.S. Federal Power Commission and U.S. Department of Energy, *Steam Electric Plant Construction Costs and Annual Production Expenses Annual Supplements*, 9 to 30, 1956 to 1977. Interestingly, the best newly installed plant for 1977 – Riverside Station – operated at 36.0% efficiency (9,476 Btu/kWh heat rate). But the best plant in service that year – the Moss Landing Station – was built in 1967 and operated at 38.2% efficiency (8,922 Btu/kWh heat rate). Also see A. F. Armor, "Power Plant Efficiency Improvement: An EPRI Perspective," in D. E. Leaver and R. G. Brown, eds., *Fossil Plant Heat Rate Improvement* (Palo Alto, CA: Electric Power Research Institute, 1981), p. 1; and "Capability," *Electrical World* 189 (March 15, 1978): 90.
9. Rankine theory incorporates the fact that water requires extra ("latent") heat to make phase changes and is affected in its changes by its pressure. Hence, the Rankine efficiency of a steam cycle diverges from the Carnot efficiency. Much of this discussion on the theory of heat engines draws from Appendix A of Paul H. Coortner and George O. G. Lof, *Water Demand for Steam Electric Generation: An Economic Projection Model* (Washington, DC: Resources for the Future, 1965 [distributed by Johns Hopkins University Press]).
10. "Conventional Steam-Electric Generating Stations," Advisory Committee Report No. 7, April 1963, Part 2, pp. 51–2 of U.S. Federal Power Commission,

National Power Survey (Washington, DC: U.S. Government Printing Office, 1964).

11. Paul L. Joskow, "Productivity Growth and Technical Change in the Generation of Electricity," *Energy Journal* 8 (January 1987): 20 (footnote 13).

12. Coortner, *Water Demand for Steam Electric Generation*, p. 18.

13. Henry Petroski, *To Engineer is Human: The Role of Failure in Successful Design* (New York: St. Martin's Press, 1985), p. 118. Good general overviews of metal failures are R. D. Barer and B. F. Peters, *Why Metals Fail* (New York: Gordon and Breach, 1970); and E. F. Bradley, ed., *Source Book on Materials for Elevated Temperature Applications* (Metals Park, OH: American Society for Metals, 1979). Also see W. R. Hibbard, Jr., "Some Problems of High-Temperature Alloys," *General Electric Review* 57 (September 1954): 57–9.

14. R. H. Richman and T. W. Rettig, eds., *Failures and Inspections of Fossil-Fired Boiler Tubes: 1983 Conference and Workshop*, EPRI CS-3272 (Palo Alto, CA: Electric Power Research Institute, 1983), p. 1–1.

15. *Ibid.*; and A. F. Armor, "Boiler Tube Failures: The Number One Availability Problem for Utilities," in *ibid.*, pp. 1–4 to 1–7.

16. F. C. Olds, "Major Rotating Equipment Design Developments," *Power Engineering* 87 (February 1983): 43.

17. Interview with a Consolidated Edison Company manager.

18. Armor, "Boiler Tube Failures," p. 1–8. See also E. Mai and G. Drucks, "Improving Reliability of Power Boilers," *Power Engineering* 83 (January 1979): 37–41. A more positive view toward solving boiler tube problems is presented in Jon Cohen, "Winning the Fight Against Boiler Tube Failure," *EPRI Journal* 11 (December 1986): 32–7.

19. Telephone interview with Louis Coffin, General Electric metallurgical engineer, November 12, 1982. A similar view was offered during an interview with S. David Freeman, director, TVA, January 17, 1984.

20. Coortner, *Water Demand for Steam Electric Generation*, pp. 19–21. A study carried out by Westinghouse Electric in 1976 came to a similar conclusion about the trade-offs between thermal efficiency and capital costs. Increasing the steam temperature in a 500-MW plant from 1,000 °F to 1,200 °F would improve thermal efficiency by 6%. However, the plant's capital cost would increase 26%. Study data cited in Robert H. Williams and Eric D. Larson, "Aeroderivative Turbines for Stationary Power," *Annual Review of Energy* 13 (1988): 441. This article also includes an excellent graph (on page 440) comparing the maximum allowable stress for different alloys used in steam tubes.

21. Interview with Walter L. Marshall, manager, Turbine Technology Laboratory, General Electric Company, Schenectady, NY, May 22, 1984.

22. Interview with a Consolidated Edison Company manager.

23. "Conventional Steam-Electric Generating Plants," p. 68.

24. "The Retreat from 1,100 F Steam," *Electrical World* 154 (October 3, 1960): 33.

25. "The Retreat from 1,100 F Takes Effect," *Electrical World* 164 (October 25, 1965): 5. Sporn noted in 1964 that since "metallurgical research had not yielded low-priced, high-temperature materials suitable for large-scale installation above 1,050 °F," AEP's next plants would use steam at a maximum temperature of 1,050 °F. Philip Sporn, "Prospects for Cost Reductions," speech given at the World Power

Conference, September 17, 1964, in Philip Sporn, *Vistas in Electric Power* Vol. 1 (Oxford: Pergamon Press, 1968), p. 199.

26. National Association of Regulatory Utility Commissioners, *1972 Report on the Ad Hoc Committee on Energy Research and Development* (Washington, DC: NARUC, 1972), included in U.S. Congress, Senate, Committee on Commerce, *Energy Research and Development II*, 93rd Congress, 1st Session, Serial No. 93–22, March 1, 1973, p. 128. A similar conclusion concerning the practical limits of thermal efficiency in conventional plants is made in D. Q. Hoover, "The Cost of Inefficiency in the Generation of Electricity," in David Japikse, ed., *The Cost of Inefficiency in Fluid Machinery* (New York: American Society of Mechanical Engineers, 1975), pp. 13–17.

27. Interview with G. O. Wessenauer, former TVA manager of Power, January 19, 1984.

28. Rajat K. Deb and James J. Mulvaney, "Economic and Engineering Factors Affecting Generating Unit Size," *IEEE Transactions on Power Apparatus and Systems* PAS-101 (October 1982): 3907.

29. "18th Steam Station Design Survey," *Electrical World* 198 (November 1984): 50.

30. Projections of energy use ten years into the future did not decline to 4.0% per year until 1980. They declined further – to 2.2% per year – in 1986. See Office of Technology Assessment, *New Electric Power Technologies: Problems and Prospects for the 1990s*, OTA-E-246 (Washington, DC: U.S. Congress, Office of Technology Assessment, 1985), p. 45. The 1986-to-1995 projection is cited in "NERC Forecasts Growth Rates," *Electrical World* 200 (July 1986): 25.

31. "Conventional Steam-Electric Generating Stations," p. 53.

32. These data were obtained by the Edison Electric Institute, an industry trade organization, and the Federal Power Commission. See John H. DeYoung, Jr. and John E. Tilton, *Public Policy and the Diffusion of Technology: An International Comparison of Large Fossil-Fueled Generating Units* (University Park, PA: Pennsylvania State University Press, 1978), p. 12. Also see C. M. Davis, K. H. Haller, and M. Wiener, "Large Utility Boilers – Experience and Design Trends," *Proceedings of the American Power Conference* 38 (1976): 281, which presents Edison Electric Institute availability data and shows the correlation between forced-outage rates and availabilities for different ranges of unit sizes. Between 1965 and 1974, units larger than 600 MW had availabilities of approximately 82%, which compared to the 90% availability for plants in the 100-to-200-MW range.

33. Verne W. Loose and Theresa Flaim, "Economies of Scale and Reliability: The Economics of Large Versus Small Generating Units," *Energy Systems and Policy* 4 (1980): 43; Lon W. Montgomery, "Large Turbine Generators," Paper No. 38, presented to the 1979 Electric Utility Engineering Conference, April 29 to May 11, 1979, p. 3; "Unit Size Vs. Availability: Is Bigger Better?" *Electric Light and Power* 55 (June 1978): 15; and "Forced Outage Rate Trends Up with Fossil Unit Size," *Electrical World* 175 (January 1, 1971): 49. Also see Paul L. Joskow, "Productivity Growth and Technical Change in the Generation of Electricity," preliminary draft of paper submitted for publication, December 31, 1985. Like the position presented in this book, Joskow concludes that the "exhaustion in scale economies is primarily a result of the poor reliability of large units. Thus, while there were clearly some cost side benefits resulting from the gradual increase in the size of generating units . . . , they were probably exhausted by about 1970." *Ibid.*, p. 33.

34. "Conventional Steam-Electric Generating Stations," p. 53.
35. Public scrutiny of the environmental effects of power production motivated the addition of auxiliary antipollution systems to standard fossil units. These auxiliaries increased the complexity of plants and their operation.
36. Robert J. Gordon, "The Productivity Slowdown in the Steam-Electric Generating Industry," unpublished paper, Northwestern University and National Bureau of Economic Research, October 1982, revised February 1983, pp. 55–6.
37. Hughes, "Scale Frontiers in Electric Power," p. 63.
38. Paul L. Joskow and Nancy L. Rose, "The Effects of Technological Change, Experience and Environmental Regulation on the Construction Costs of Coal-Burning Generating Units," MIT Working Paper No. 336, Department of Economics, MIT, March 1984, p. 5.
39. *Ibid.*, p. 6. The move away from supercritical plants is also described in H. B. Finger, "Electric Power Plant Efficiency," in Brian J. Gallagher, ed., *Energy Production and Thermal Effects* (Ann Arbor, MI: Ann Arbor Science Publishers, 1974), pp. 130–8. General Electric concluded that reliability problems were related more to the large size of supercritical units than to their use of steam at high temperatures and pressures. R. C. Spencer, "Design of Double-Reheat Turbines for Supercritical Pressures," *Proceedings of the American Power Conference* 42 (1980): 226.
40. Interview with Christopher C. Rolfe, manager, Research and Projects, Design Engineering, Duke Power Company, August 8, 1985.
41. Interview with George H. Kimmons, manager of Engineering Design and Construction, TVA, January 17, 1984.
42. "Toward Simplicity in Nuclear Plant Design," *EPRI Journal* 11 (July/August 1986): 5.
43. Richard Davies, *Learning by Doing and Using and Electric Utility Input into the Nuclear Powerplant Innovation Process*, M.S. thesis, Technology and Policy, Massachusetts Institute of Technology, May 1984, p. 100.
44. Westinghouse Electric Corporation, "Turbine Components Plant – Winston-Salem, NC," Westinghouse promotional pamphlet, no date, p. 9.
45. See Richard F. Hirsh, "Westinghouse Electric Corporation – Steam Turbine Division (A)," Case 0–687–036, *Harvard Business School Case Studies Series* (1986).
46. Interview with Norman H. Jones, manager, General Electric Company, Generator Engineering Section, May 23, 1984. Also see GE news releases, February 1, 1978 and November 8, 1979. Westinghouse Electric Corporation and a research group at MIT, along with competitors in Japan, France, West Germany and the United States, also worked on large superconducting generators in the 1970s and 1980s. A superconducting material is one in which electricity passes with no resistance. For a review of this work, see John S. Joyce, Wilhelm Engelke, and Dietrich Lambrecht, "Will Large Turbine-Generators of the Future Require Superconducting-Field Turbogenerators?" *Proceedings of the American Power Conference* 39 (1977): 255–69; and J. S. Edmonds and W. R. McCown, "Large Superconducting Generators for Electric Utility Applications – The Prospects," *Proceedings of the American Power Conference* 42 (1980): 629–38. Also see E. Pannen, "Grenzen im Turbogeneratorenbau," *Elektrizitaetswirtschaft* 75 (1976): 682–90. An update on superconductor technology following discoveries of "hot" superconduc-

tors is John Douglas, "Pursuing the Promise of Superconductivity," *EPRI Journal* 12 (September 1987): 5–15.

47. "Energy-cost Record May be Hard to Beat," *Electrical World* 168 (October 16, 1967): 73.

48. *Ibid.*

49. *Ibid.*

50. Interview with Hugh J. Murphy, general manager, Large Steam Turbine-Generator Marketing Operations, General Electric Company, May 21, 1984. Murphy has an undergraduate degree in metallurgical engineering and worked as a process engineer at the GE Knolls Atomic Power Laboratory.

51. The radiation makes pipes more brittle than engineers foresaw when they planned nuclear units, meaning that many plants will need to be retired before originally expected. See Lindsey Gruson, "Nuclear Power Plant Dismantled," *The New York Times*, "Science Times" section (November 25, 1986), pp. C1 and C3.

52. Petroski, *To Engineer is Human*, p. 118. Deterioration of tubes in nuclear steam generators is discussed in Bob Pollard, "PWR Steam Generator Tube Deterioration: Another Nuclear Power Problem That Won't Fade Away," *Nucleus* 4 (Summer 1982): 7–8. *Nucleus* is a publication of the Union of Concerned Scientists.

53. Davies, "Learning by Doing," p. 30.

54. Richard Schmalensee, MIT, phone call to author, March 1984.

55. Emphasis added. Gordon, "The Productivity Slowdown," p. 65.

Chapter 8

1. The calculation leading to this figure is similar to one performed by Philip Sporn in 1970. Looking at the power industry's capacity needs, Sporn estimated that an 8% annual load growth would mean that orders should have been placed for 104,000 MW from 1965 through 1969, or about 21,000 MW per year. Philip Sporn, "Developments in Nuclear Power Economics: January 1968–December 1969," in U.S. Congress, Joint Committee on Atomic Energy, *Nuclear Power and Related Energy Problems – 1968 through 1970*, 92nd Congress, 1st Session, December 1971, pp. 2 and 15. Data on installed capacity comes from Edison Electric Institute, *Edison Electric Institute Pocketbook of Electric Utility Statistics* (Washington, DC: Edison Electric Institute, 1984), p. 5.

2. Bruce T. Allen and Arie Melnik, *The Market for Electrical Generating Equipment* (East Lansing, MI: MSU Public Utilities Papers, 1973), p. 49.

3. The ordering cycles of utilities are discussed in Jules Backman, *The Economics of the Electrical Machinery Industry* (New York: New York University Press, 1962), p. 131.

4. "Allis-Chalmers to Stop Making Generators," *Wall Street Journal* (December 21, 1962), p. 32; "Allis-Chalmers Steam Turbine Exit Shocks Industry," *Electrical World* 158 (December 24, 1962): 22–3; and "One Less Steam Turbine Supplier," *Electrical World* 159 (January 7, 1963): 14. The company returned to the power business in 1970 in a joint venture with West Germany's Kraftwerk Union. "Out of the Shadow of 'Big Allis,' " *Business Week*, No. 2186 (July 24, 1971): 22.

5. Ralph G. M. Sultan, *Pricing in the Electrical Oligopoly*, Vol. 2 (Cambridge, MA: Harvard University Press, 1975), p. 87.

6. Neither government nor industry planners anticipated this high growth rate, which occurred in 1966, 1968, and 1969. U.S. Congress, Senate, Committee on

Commerce, *Background Report on Powerplant Siting*, 92nd Congress, 2nd Session, July 1972, p. 1; included in U.S. Congress, Senate, Committee on Commerce, *Powerplant Siting*, 92nd Congress, 2nd Session, Serial No. 92–92, April 28, May 15, and June 1, 1972, p. 714.

7. Allen, *The Market for Electrical Generating Equipment*, pp. 49–50.

8. "Power Happy," *The Economist* 225 (December 23, 1967): 1228.

9. Sultan, *Pricing in the Electrical Oligopoly*, Vol. 2, p. 163. An *Electrical World* editorial writer tried to sound an alarm in 1966, noting that the previous boom-bust cycle that peaked in 1957 led to the loss of one turbine-generator manufacturer (Allis-Chalmers) and the reduction of research-and-development expenditures by the remaining two American vendors. Utilities should not be in such a hurry to buy new equipment, noted the writer. "What's the Big Rush?" *Electrical World* 165 (February 21, 1966): 7.

10. "New Generating Capacity Plans Pegged at 137.2 Million KW by '74," *Electrical World* 167 (February 20, 1967): 70.

11. " 'Fed Up' Over Price Policy, AEP Buys Turbines Abroad," *Electrical World* 169 (January 1, 1968): 15; and Arthur J. Stegeman, "T-G Competition is More Than Mere Numbers," *Electric Light and Power* 46 (January 1968): 5. The offshore purchase spurred General Electric to insert a letter in the May 1968 issue of *Electric Light and Power* (Volume 46, between pages 162 and 163) indicating that its turbine generators outperformed those of foreign competitors, who received government subsidies to compensate for lower prices.

12. " 'Fed Up,' " p. 15.

13. "The Turbine Troubles that Plague Westinghouse," *Business Week*, No. 2325 (April 6, 1974): 54–5.

14. Interview with John E. Dolan, vice chairman, Engineering and Construction, AEP Service Corporation, July 30, 1985.

15. The desire for lower unit costs created some tension between the manufacturers and their customers. Utilities argued that manufacturers concerned themselves only with high profits, whereas the vendors pointed to their need for large investments in new production capacity. In a 1968 editorial entitled "This Believability Gap Must Be Closed," *Electrical World* noted that manufacturers believed that "[p]rice is the sole source of utility dissatisfaction at a time when reliability and efficiency of United States systems should be paramount." "This Believability Gap Must Be Closed," *Electrical World* 169 (January 8, 1968): 15. Defending the manufacturers, Robert A. Lincicome, managing editor of an industry journal, titled an editorial "Methinks Thou Dost Protest Too Loudly," *Electric Light and Power* 49 (April 1971): 3. In 1973, a manufacturing manager repeated the view that in the current situation, vendors could do little but try to reduce the cost of production, even when quality suffered. As argued by GE's J. T. Peters, utilities would only be able to meet the increasing demands of their customers if they "recognize and support products that are superior in operational ability and longevity. Quality and performance competition will grow *if the purchasers so insist.*" Emphasis added. J. T. Peters, "That Vital Commodity, Quality," *Public Utilities Fortnightly* 91 (May 10, 1973): 33.

16. These data were derived from a graph presented in "18th Steam Station Design Survey," *Electrical World* 198 (November 1984): 50.

17. R. J. Niebo, "Power Plant Productivity Trends and Improvement Possibilities," *Combustion* 50 (May 1979): 35.

18. Gaining experience with the technology would have been difficult anyway due to a statistical reason: as power units "grew" in size, utilities needed fewer of them to meet demand for electricity and thus, utilities and manufacturers had less chance to gain field experience. For example, in 1949, utilities installed 162 new fossil-fueled steam units in an attempt to catch up with demand created during World War II. But since then, utilities built fewer and fewer units, with 77 (and 2 nuclear) units installed in 1960 and only 56 (and 4 nuclear) installed in 1970. Data from U.S. Department of Energy, Energy Information Administration, Generating Unit Reference File, February 24, 1987. The effect of the trend was that manufacturers simply were unable to gain as much operational experience with their equipment as they had in the immediate postwar era. Less experience was simply a corollary of unit size increasing at a rapid rate. It was also one reason that contributed to Allis-Chalmer's departure from the turbine business. As the number 3 producer in a three-company oligopoly, Allis-Chalmers was the least likely to get many orders in a business that saw fewer but larger units. See "One Less Steam Turbine Supplier," p. 14.

19. Interview with a Consolidated Edison Company manager.

20. Linda Charlton, "Con Ed Loses 2d Big Generator for Summer," *The New York Times*, July 23, 1970, p. 1. The problems with "Big Allis" are recounted in Alexander Lurkis, *The Power Brink: Con Edison – A Centennial of Electricity* (New York: Icare Press, 1982), pp. 109–12.

21. Interview with a Consolidated Edison Company manager.

22. Jeremy Main, "A Peak Load of Trouble for the Utilities," *Fortune* 80 (November 1969): 194.

23. "Why Utilities Can't Meet Demand," *Business Week*, No. 2100 (November 29, 1969): 49–50.

24. Henry C. Heisler, "Warranties: Who Needs Them?" *Electric Light and Power* 48 (February 1970): 64.

25. Philip Sporn, *The Social Organization of Electric Power in Modern Societies* (Cambridge, MA: The MIT Press, 1971), p. 105.

26. Electric Research Council, *Electric Utilities Industry Research and Development Goals Through the Year 2000: Report of the R&D Goals Task Force to the Electric Research Council*, ERC Publication No. 1–71 (New York: Electric Research Council, 1971), p. 15.

27. "Conventional Steam-Electric Generating Stations," Advisory Committee Report No. 7, April 1963, Part 2, p. 54 of U.S. Federal Power Commission, *National Power Survey* (Washington, DC: U.S. Government Printing Office, 1964).

28. Interview with Eugene J. Cattabiani, executive vice president, Power Generation, Westinghouse Electric Corporation, July 29, 1985.

29. Interview with John F. Traexler, manager of Engineering, Steam Turbine Generating Division, Westinghouse Electric Corporation, April 25, 1986.

30. *Ibid*.

31. Letter from Eugene J. Cattabiani, December 13, 1988, and interview with Mr. Cattabiani, July 29, 1985.

32. *Ibid*.

33. Peters, "That Vital Commodity, Quality," p. 34.

34. The next largest chunk of problems occurred from maintenance shortcomings, a problem caused by utilities. *Ibid*.

35. *Ibid*., pp. 35–6.

36. Interview with George B. Cox, senior vice president and group executive, Turbine Business Group, General Electric Company, May 21, 1984. To add insult to injury, the American Electric Power Company sued GE and Westinghouse in 1971 for monopolizing the turbine-generator business through price fixing. According to the AEP, General Electric established prices in its price book, which was duplicated exactly by Westinghouse. In 1977, the manufacturers agreed to modify earlier price-fixing consent decrees that supposedly eliminated the price book similarities, resulting in the end of the suit. Interview with John E. Dolan, AEP, July 30, 1985; AEP, *Annual Report, 1971*, p. 35; and American Electric Power Company, Annual Report, Form U5S filed with the Securities and Exchange Commission, File No. 30–150, for the year ended December 31, 1977. This document summarizes the result of litigations during the year. See also "Turbine Makers Get Sued Again," *Business Week*, No. 2210 (January 8, 1972): 24.
37. Peters, "That Vital Commodity," p. 37.
38. U.S. Atomic Energy Commission, "Regulatory Guide No. 1.49, Power Levels of Nuclear Power Plants," December 1973, p. 1.
39. Emphasis added in *ibid*.
40. Interview with Dwight Patterson, assistant to the TVA manager of Engineering Design, January 18, 1984.
41. Interview with John F. Traexler.
42. The relationship between manufacturers and utilities became increasingly hostile within an unusual business environment that emerged in the 1960s. After a highly cyclical (and not very profitable) period of new equipment orders in the 1950s, the early 1960s witnessed hearty price competition after the federal government prosecuted vendors for engaging in collusive activities. With several years occupied by courtroom battles, manufacturers of power equipment (which included other companies besides just General Electric, Westinghouse, and Allis-Chalmers) had suits filed against them that could have resulted in $9 billion in treble damages. In the end, the manufacturers paid damages that observers estimated could have exceeded $600 million. For accounts of the collusion and trials, see Sultan, *Pricing in the Electrical Oligopoly*, Vol. 1.; Clarence C. Walton and Frederick W. Cleveland, Jr., *Corporations on Trial: The Electric Cases* (Belmont, CA: Wadsworth, 1964); John Herling, *The Great Price Conspiracy: The Story of the Antitrust Violations in the Electrical Industry* (Washington, DC: Robert B. Luce, 1962); and "Antitrust and the Organization Man," *Wall Street Journal* (January 10, 1961), p. 10.
43. *AEP, Annual Report, 1974*, p. 34.
44. "The Turbine Troubles that Plague Westinghouse," p. 54. Also see the discussion of suits against GE and Westinghouse in the AEP *Annual Report*, 1974.
45. "The Turbine Troubles that Plague Westinghouse," p. 55.
46. Interview with E. Floyd Thomas, retired TVA manager of Power Operations, January 19, 1984.
47. Interview with Hugh G. Parris, TVA manager of Power, January 19, 1984.
48. Interview with S. David Freeman, TVA director, January 17, 1984.
49. Interview with Bide L. Thomas, executive vice president, Commonwealth Edison Company, August 20, 1985, and letter from Mr. Thomas, January 24, 1989.
50. Similar experiences of engineers making fundamental errors when using computers are examined in Henry Petroski, *To Engineer is Human: The Role of Failure in Successful Design* (New York: St. Martin's Press, 1985).
51. Interview with John F. Traexler.

Chapter 9

1. See *Economic Report of the President, 1980* (Washington, DC: U.S. Government Printing Office, 1980), pp. 205, 207, 208, and 278; cited in Thomas K. McCraw, *Prophets of Regulation* (Cambridge, MA: Harvard University Press, 1984), p. 237.

2. Interview with Samuel C. Brown, Jr., senior vice president, Power Station and Construction, Virginia Electric and Power Company, August 20, 1981.

3. Leonard S. Hyman, *America's Electric Utilities: Past, Present, and Future* (Arlington, VA: Public Utilities Reports, 1983), pp. 90 and 110.

4. The overall average yield of forty utility bonds, ten each in the Moody's ratings of Aaa, Aa, A, and Baa, fell to 8.46% in June 1977. Yields peaked in September 1981 at 17.56%. Edison Electric Institute, *Statistical Yearbook of the Electric Utility Industry/1982* (Washington, DC: Edison Electric Institute, 1983), p. 85.

5. Address given by William Derrickson, vice president for Nuclear Power, Public Service Company of New Hampshire, before a group of MIT students, February 12, 1985.

6. Richard K. Lester, "Rethinking Nuclear Power," *Scientific American* 254 (March 1986): 32.

7. "Coal Strap Tightens More in Southeast," *Electrical World* 173 (April 27, 1970): 22–3; "TVA Says Coal Supply May Run Out, Causing a Shortage of Power," *Wall Street Journal* (August 20, 1970), p. 28; and "American Electric Says Its Supplies of Coal are 'Dangerously Low,' " *Wall Street Journal* (September 4, 1970), p. 8.

8. Richard H. K. Vietor, *Energy Policy in America Since 1945: A Study of Business-Government Relations* (Cambridge: Cambridge University Press, 1984), p. 193.

9. Hyman, *America's Electric Utilities*, p. 100. Overall, fuel costs increased from a low (in nominal terms) of $6.22 in 1966 to $9.31 in 1972 per ton of coal equivalent. But a bigger jump in fuel costs occurred after the beginning of the Arab oil embargo in the winter of 1973, when costs jumped from $10.65 in 1973 to $18.49 the next year, and $37.60 by 1980 for each ton of coal equivalent. Data from Edison Electric Institute, *Edison Electric Institute Pocketbook of Electric Utility Statistics*, 30th Ed. (Washington, DC: Edison Electric Institute, 1984), p. 22.

10. The price of coal did not increase as much as oil, but it did rise substantially. Bituminous coal prices rose 212% from 1970 to 1980. Hyman, *America's Electric Utilities*, p. 101. Even so, in 1982, the average cost of a Btu of energy from coal used by utilities was 35% the cost of oil. Edison Electric Institute, *Statistical Yearbook/1982*, p. 36.

11. In 1970, sulfur-dioxide removal was expected to cost $25 to $50 per kilowatt at a time when the incremental cost of capacity was $147/kW – well over 10%. 1970 estimates for the aggregated capital costs of building new facilities to cool water before it was discharged ranged from $2.7 to $4.8 billion (1970 dollars) in 1980. ($800 million was spent in 1970 for this purpose.) Meanwhile, the cost of cooling the water in 1980 was expected to increase electric energy costs by 0.22 mill to 0.39 mill/kWh compared to the 1970 increase of 0.15 mill/kWh. Federal Power Commission, *The 1970 National Power Survey*, Part 1 (Washington, DC: U.S. Government Printing Office, 1971), pp. I-22–6 and I-10–20. Later estimates of the cost to meet environmental regulations reaffirmed the view that the task would be expensive.

For oil-fired generating plants alone, it was expected that companies would have to invest 15.4% more capital to meet 1976 air quality standards than when regulations were less stringent. Carolyn H. Brancato, "A Breakthrough in Electricity Pricing," *New York Affairs* 1 (1974): 81. As an example of actual costs, consider Pennsylvania Power Company's Mansfield plant, completed in 1976, which included "scrubbers" for removing 92% of the sulfur dioxide from coal smoke. At a cost of $400 million for a plant that included three 800-MW units, the environmental equipment cost $167 per kilowatt. The per-kilowatt cost of the entire plant amounted to $452, making the add-on machinery a significant portion of the plant's total cost. The cost of the scrubbers is included in Jerry Ackerman, "Scrubbers Could Curb Acid Rain," *Boston Globe* (February 20, 1984), pp. 37 and 40. Other data concerning the Mansfield plant were found in U.S. Department of Energy, Energy Information Administration, *Thermal-Electric Plant Construction Cost and Annual Production Expenses–1980* (Washington, DC: U.S. Government Printing Office, 1983), p. 137. For more on the effects of the 1970 Clean Air Act and the 1977 Amendments, see Peter Navarro, "The Politics of Air Pollution," *Public Interest* No. 59 (Spring 1980): 36–44.

12. One econometric study on the effect of reducing sulfur-dioxide emissions found that 44% of the overall productivity decrease in the period from 1973 to 1979 could be attributed to compliance of the new environmental regulations. Productivity declined for firms facing the binding emission constraints an average of 1.35% per year during 1973–9, of which a decline of 0.59% resulted from regulation alone. Frank M. Gollop and Mark J. Roberts, "Environmental Regulations and Productivity Growth: The Case of Fossil-fueled Electric Power Generation," *Journal of Political Economy* 91 (1983): 654–74.

13. Interview with Samuel C. Brown, Jr.; Virginia Electric and Power Company, *Annual Reports* for 1961, 1970, and 1980. The effect of environmental control equipment on heat rates of power plants is discussed briefly in U.S. Department of Energy, *Thermal-Electric Plant Construction Cost and Annual Production Expenses–1980*, p. 10. And in a 1985 article in *Science*, it was pointed out that one benefit of using a new integrated gasification combined-cycle technology was to avoid the 10% capacity penalty suffered by utilities as a result of using conventional flue-gas scrubbers. Mark Crawford, "Utilities Look to New Coal Combustion Technology," *Science* 228 (May 3, 1985): 565.

14. Charles F. Phillips, Jr., *The Regulation of Public Utilities: Theory and Practice* (Arlington, VA: Public Utilities Reports, 1984), pp. 19–20. Suppression of rate increases is also discussed in Peter Navarro, "The Performance of Utility Commissions," in John C. Moorhouse, ed., *Electric Power: Deregulation and the Public Interest* (San Francisco: Pacific Research Institute for Public Policy, 1986), pp. 339–42.

15. Moody's Investor Service, *Moody's Public Utility Manual* (New York: Moody's Investor Service, 1982), p. a-11.

16. A summary of the state and federal regulations that utilities needed to meet in the 1960s through the 1980s is included in A. H. Ringleb, "Environmental Regulation of Electric Utilities," in Moorhouse, ed., *Electric Power: Deregulation and the Public Interest*, pp. 183–218.

17. U.S. Department of Energy, *Impacts of Financial Constraints on the Electric Utility Industry*, DOE/EIA-0311 (Washington, DC: U.S. Government Printing Office, 1981), pp. A-12 and A-21. Also see Appendix C of this document for a listing of the few states that permitted CWIP in the rate base as of 1981.

18. The Federal Power Commission studied the cause of delays in 114 units having capacities greater than 300 MW between 1966 and 1970. It found 52% involved labor problems; 23% could be attributed to equipment failures, faulty installation, and start-up problems; 14% was due to late delivery of equipment; and 6% resulted from regulatory (including environmental) concerns. U.S. Congress, Senate, Committee on Commerce, *Background Report on Powerplant Siting*, 92nd Congress, 2nd Session, July 1972, p. 11; included in U.S. Congress, Senate, Committee on Commerce, *Powerplant Siting*, 92nd Congress, 2nd session, Serial No. 92–92, April 28, May 15, and June 1, 1972, p. 724.

19. "Why Utilities Can't Meet Demand," *Business Week* No. 2100 (November 29, 1969): 48.

20. The long and unsuccessful attempt by Consolidated Edison to build its Storm King plant is recounted in Allan R. Talbot, *Power Along the Hudson: The Storm King Case and the Birth of Environmentalism* (New York: Dutton, 1972).

21. Thomas P. Hughes, *Networks of Power: Electrification in Western Society, 1880–1930* (Baltimore: Johns Hopkins University Press, 1983), pp. 142–6.

22. *Ibid*., p. 173.

23. *Ibid*., pp. 385–90.

24. Interview with Peter D. Hindley, senior mechanical engineer, Generation Planning Department, Pacific Gas and Electric Company, July 8, 1982.

25. Though not part of the utility industry, the early manufacturers of power equipment resembled the giants of high tech today, such as IBM, Hewlett-Packard, and Xerox, who spent heavily on research and development to consolidate their "first-mover" status in the field. In fact, General Electric became one of the first corporations to build an industrial research laboratory. Its managers realized that the new technology required continuous innovation in order to maintain leadership in a field that they pioneered. See George Wise, *Willis R. Whitney, General Electric, and the Origins of United States Industrial Research* (New York: Columbia University Press, 1985); and Leonard S. Reich, *The Making of American Industrial Research: Science and Business at GE and Bell, 1876–1926* (New York: Cambridge University Press, 1985).

26. In the 1930s, it took four or five years of gross revenues to equal the capital invested. J. V. Garland and Charles F. Phillips, *The Crisis in the Electric Utilities,* Vol. 10, No. 10, in *The Reference Shelf* series (New York: H. W. Wilson, 1936), p. 31. This figure did not change much by 1970, when $4.71 of capital investment produced $1 of income. Bruce A. Smith, *Technological Innovation in Electric Power Generation: 1950–1970* (East Lansing, MI: Michigan State University, 1977), pp. 6–7.

27. Interview with Christopher C. Rolfe, manager, Research and Projects, Design Engineering, August 8, 1985; and letter from Mr. Rolfe, January 19, 1989.

28. Exceptions included executives at the Detroit Edison, Commonwealth Edison, and Pacific Gas and Electric companies, who sought approval from the Atomic Energy Commission to work with constructors on designs for nuclear power plants in the early 1950s. George T. Mazuzan and J. Samuel Walker, *Controlling the Atom: The Beginnings of Nuclear Regulation, 1946–1962* (Berkeley: University of California Press, 1984), pp. 18–19.

29. The vendors often lost a great deal of money building turn-key plants. The experience of General Electric provides an instructive example. The company decided to construct the first commercial-size nuclear plant, Dresden I (200 MW), in 1955 (completed in 1960). Though the firm knew it would lose $15 to $20 million on the deal, it hoped to win market share as new orders poured in. The company waited glumly for several years, as utilities refused to purchase the GE design, while going, instead, to larger plants designed by Westinghouse. Hoping to "ram this thing right on through," CEO Fred J. Borch offered to build nuclear plants on a turn-key basis, in which the company supplied all the components and took full responsibility for building the entire plant at a fixed price. GE subsequently won plenty of orders. Between 1963 and 1966, it wrote sixteen American and three foreign contracts, ten of them on a turn-key basis. Business was greater than expected, but because the company wrote off engineering and development costs as they were incurred, the fast growth meant tremendous financial losses. Compounding this financial problem were turn-key contracts that consistently lost money for the company. Moreover, the company's lack of experience in building nuclear plants impeded its financial health. As utilities demanded larger plants, GE could only estimate – based on little experience – how much the plants would cost. Expecting low inflation and moderate construction cost increases over the years, the company suffered as construction and labor costs soared as much as 20% per year in some areas. GE also underestimated the amount of construction labor by as much as 50%. In short, the "orders were coming so fast that GE was accepting them before it had any experience with the vicissitudes of construction." The experience caused GE to leave the turn-key business (with Westinghouse following suit) with a loss of over $200 million. See Allan T. Demaree, "GE's Costly Ventures into the Future," *Fortune* 87 (October 1970): 93. The use of "loss-leader" pricing helped American companies win several orders in Europe – especially France. See Robert L. Frost, "France's 'False Start' in Nuclear Power, 1954–1969: The Death of Independent French Nuclear Technology," paper presented at the annual meeting of the Society for the History of Technology, Cambridge, MA, November 1–4, 1984.

30. The Department of Energy estimated that government provided subsidies totalling about $37 billion to the nuclear industry in the three decades before 1980. John Emshwiller, "The Trouble with Utilities," *The Nation* (April 4, 1981): 393.

31. Interview with Duke Power Company engineers, August 8, 1985; and Richard H. Graham, "Nuclear Fuel-Cost Trends Under Private Ownership," *Proceedings of the American Power Conference* 27 (1965): 210–13.

32. This episode provides grist for the interesting R&D management issue of how vendors affect the market and vice versa, and how "outside," nonmarket forces affect industrial practices.

33. Philip Sporn, "Place of Nuclear Energy in United States Power Supply," paper given October 16, 1952, in Philip Sporn, *Vistas in Electric Power*, Vol. 1 (Oxford: Pergamon Press, 1968), p. 213.

34. Unlike several other companies, however, Commonwealth Edison had relatively good experiences with nuclear power in the 1960s and 1970s. Interview with John J. Viera, vice president, Commonwealth Edison Company, August 20, 1985. Duke Power Company managers indicated that they also felt initially that nuclear power would not require new technical or administrative skills. Interviews with several managers, August 8, 1985. Apparently, the view still persists among some observers. Retired Admiral Joe Williams, Jr., a protege of Hyman Rickover's nuclear navy program, feels that many managers of nuclear plants do not understand the special demands of this technology. Admiral Williams was hired by Toledo Edison after the company's Davis-Besse nuclear plant had a serious accident. See "America's Big Risk," *Newsweek* 109 (April 27, 1987): 58.

35. Quotation of James H. Campbell, president of Consumers Power Company, in "A Turn to Bits and Pieces for Nuclear," *Electrical World* 165 (April 18, 1966): 62.

36. The conservative and risk-averse nature of electric utility managers is discussed in Robert D. Brenner, "Electric Utilities as Markets for Advanced Generating Technologies," *Public Utilities Fortnightly* 101 (April 13, 1978): 25–33.

37. "Where are the Bright Engineers?" *Electrical World* 170 (July 8, 1968): 7.

38. "Recruiting: A Problem of Language?" *Electrical World* 170 (September 9, 1968): 67–70. AEP's Philip Sporn also despaired over the industry's inability to attract good students. See his speech, "Challenges of Electric Power's Future," given in 1961, in Sporn, *Vistas in Electric Power*, Vol. 1, pp. 169–70.

39. "Young Utility Engineers Take Critical Look at Power Work and Recruiting," *Electric Light and Power* 42 (January 1964): 28.

40. John D. Alden, "Manpower Trends in Electrical Engineering in the United States," *Proceedings of the IEEE* 59 (1971): 835.

41. For evidence of recruiting problems, see, for example, "Promote Utility Careers," *Electrical World* 145 (February 27, 1956): 5; "Will Utilities Get the Talent They Need?" *Electrical World* 145 (March 12, 1956): 62; and "Utilities Came Within 93.4% of Goal Fulfillment," *Electrical World* 162 (October 12, 1964): 32–3.

42. Lawrence J. Hollander, "The Big Blackout: Whooping Cranes and Power Failures," *The Nation* 202 (January 10, 1966): 34.

43. Glenn Zorpette, "Scarce Programs: Power Engineering," *IEEE Spectrum* 21 (November 1984): 54–5.

44. Frederick E. Terman, "Electrical Engineers are Going Back to Science!" *IRE Student Quarterly* 2 (December 1955): 3. A *Fortune* article in 1969 attributed the loss of good managerial talent to problems dating back to the 1930s. "When the huge utility holding companies were broken up in the 1930s," the article noted, "second-string managers were put in charge of the bits and pieces, and until lately, there has been little to challenge them and their successors. Utilities tend to recruit average students from smaller colleges, and they are usually young men of modest ambition willing to wait for the rewards of age and seniority." Jeremy Main, "A Peak Load of Trouble for the Utilities," *Fortune* 80 (November 1969): 196. The article also suggested that "utility executives are generally unimaginative men, grown complacent on private monopoly and regulated profits." *Ibid.*, p. 118.

45. "Is the Slumbering Giant Waking Up?" *Electrical World* 168 (August 14, 1967): 39.

46. "Recruiting," p. 68.

47. H. A. Cavanaugh, "The Management Report," *Electrical World* 198 (September 1984): 33.

48. This quotation comes from a conversation between the article's author and a colleague in E. W. Greenfield, "Manpower Needs in Electric Power Engineering," *Proceedings of the American Power Conference* 27 (1965): 71. It was recounted as a way to argue that the power industry truly needed more of the type of people that his colleague did not appreciate.

49. Interview with S. David Freeman, director, TVA, January 17, 1984. Also see the criticism of the industry's hiring practices and the perceived "luxury" of hiring students with graduate engineering degrees, in Hollander, "The Big Blackout," p. 34.

50. "Why Utilities Can't Meet Demand," *Business Week*, No. 2100 (November 29, 1969): 48.
51. John R. Emswiller, "Job of Managing Utility Loses its Allure, A Victim of Industry's Mounting Problems," *Wall Street Journal* (May 18, 1981), p. 29.
52. "Who Made What at the Top in U.S. Business," *Forbes* 135 (June 3, 1985): 114–6.
53. R. S. Neblett, "Steam Turbine-Generators of the Future," *General Electric Review* 58 (July 1955): 13.
54. H. J. Harlow, "Steam Plant Improvements Projected to 1980," in *Prospects for Economic Nuclear Power* (New York: NICB, 1957), p. 11, cited in General Electric, *1980: The Basic Planning Horizon*, Vol. 2, marketing services report (Schenectady, NY: General Electric Company, 1958), p. 174.
55. Had the investor-owned utility industry reached the predicted average efficiency in 1980, it would have saved about $3.7 billion, the equivalent of 19% of the year's construction expenditures for new production facilities. Figures calculated by author from data in Edison Electric Institute, *Pocketbook*, pp. 6, 15, and 21.
56. R. R. Bennett, "Station Size – Where are We Heading?" *Proceedings of the American Power Conference* 30 (1968): 488–9.
57. *Ibid.*
58. See Philip Sporn, "Inventing Our Future," talk at AEP Management Meeting, Wheeling, WV, November 10–12, 1964, p. 10.
59. Reflecting such optimism, the utility-sponsored Electric Research Council set 2,500 MW as a goal for new turbine-generator sets by the mid-1980s. See Electric Research Council, *Electric Utility Industry Research and Development Goals through the Year 2000: Report of the R&D Goals Task Force to the Electric Research Council* (New York: Electric Research Council, 1971), p. 15.
60. Edwin Vennard, *Management of the Electric Energy Business* (New York: McGraw-Hill, 1979), p. 103.
61. For a contemporaneous and optimistic assessment of nuclear power, see John F. Hogerton, "The Arrival of Nuclear Power," *Scientific American* 218 (February 1968): 21–31.
62. The plans are described in U.S. Congress, Senate, Committee on Commerce, *Powerplant Siting*, 92nd Congress, 2nd session, Serial No. 92–92, April 28, May 15, and June 1, 1972, pp. 966–1026.
63. As summarized by a special task force convened by the Federal Power Commission in 1970, "special attention [should] be directed to nuclear power because it promises the best available answer to many of the existing environmental problems." Task Force on Environment, "Managing the Power Supply and the Environment," Report to the Federal Power Commission, July 1, 1971, in Federal Power Commission, *The 1970 National Power Survey*, Part 1 (Washington, DC: U.S. Government Printing Office, 1971), p. I-22–11. The task force had a diverse membership, including S. David Freeman, who was later a critic of nuclear power, then serving as Energy Policy Director of the President's Office of Science and Technology. Other members included utility managers, an Environmental Protection Agency commissioner, an Atomic Energy Commission director, a university environmental engineer, and a sport fishermen's representative.
64. Irvin C. Bupp and Jean-Claude Derian, *The Failed Promise of Nuclear Power: The Story of Light Water*, Chapter 5, "High Tide for Light Water" (New York: Basic Books, 1981). Moreover, nuclear plants conserved coal and oil, and they obviated

the need to transport bulky fuels. They would even save nuclear fuels once breeder reactors came on line. Offering the advantage of using the waste of other plants, these reactors would effectively deal with the difficulty of disposing radioactive by-products, one of the few apparent shortcomings of the technology. In fact, the breeder reactors had become the "cause celebre" of the power industry, with the Electric Research Council calling it in 1971 a "critically important" research goal for commercial availability by the mid-1980s. Electric Research Council, *Electric Utilities Industry Research and Development Goals*, p. 11. Reflecting the enthusiasm, President Nixon called it "our best hope today for meeting the Nation's growing demand for economical clean energy." Richard M. Nixon, "Proposals for Adequate Supply of Clean Energy," message to Congress sent June 4, 1971, contained in Richard M. Nixon, *Nixon: The Third Year of His Presidency* (Washington, DC: Congressional Quarterly, 1972), p. 91-A.

65. This quote comes from Eugene Ferguson, one of the history of technology's "deans," in a letter to the author, January 14, 1985.

66. This was one reason cited by observers for the failure of many utilities to build and manage effectively their nuclear power plants. See John Tirman, "Nuclear Industry, Heal Thyself," *Wall Street Journal* (December 30, 1983), p. 8. Mr. Tirman wrote as senior editor for the Union of Concerned Scientists.

67. Interview of Hugh G. Parris, TVA manager of Power, January 19, 1984.

68. Interview with Thomas H. McCloskey and A. F. Armor, Electric Power Research Institute, July 12, 1982.

69. Interview with Dwight Patterson, assistant to the TVA manager of Engineering Design, January 18, 1984.

70. While many TVA managers wanted to believe that they could extrapolate the sizes of both fossil and nuclear units without encountering problems, others within the organization had their reservations. E. Floyd Thomas, former TVA manager of Power Operations, for example, pointed out to his colleagues that problems had already occurred when General Electric extrapolated a 60-MW unit to 100 MW. Multiple failures with the nozzle and first-stage blades occurred before trial-and-error repairs and design changes corrected the problems. The jump from 150- to 300-MW units did not cause many problems, but "the 300 to 500 MW jump was again a major problem requiring many emergency outages and major modification work." Similar extrapolation problems existed with the 900-, 1,100-, and 1,300-MW fossil units. As Mr. Thomas put it in a letter, TVA was "going from one extrapolated size to another without allowing time to factor in lessons learned from previous units. We should have leveled off at 500- to 600-MW units." When TVA and other utilities went to nuclear, they followed a similar philosophy in order to reduce costs. Unfortunately, problems with nuclear units could not be resolved as easily as fossil machines since repair work and modifications on nuclear plants required regulatory oversight and approval. Letter from E. Floyd Thomas to author, December 10, 1988. Other comments came during an interview with Mr. Thomas, January 19, 1984.

71. The average residential rate jumped from 0.89 to 0.93 cents per kilowatt-hour, still well under half the average rate for the nation of 2.14 cents. TVA customers also used much more electricity than the average American – 12,668 kWh in 1967 versus 5,788 for the average American. TVA, *Power Annual Report, 1968*, pp. 405.

72. "Rate Actions, Fuel Clauses Help to Finance Plant Expansion," *Electrical World* 181 (March 15, 1974): 60.

73. *Ibid*. Utilities actually requested $455 million by August 1969, though many requests had not yet been acted upon. "Higher Rates Sought in Face of Cost Rises," *Electrical World* 172 (September 1, 1969): 17.
74. Edison Electric Institute, *Pocketbook*, p. 22.
75. "Increase in Sales, Higher Rates Help Restore Industry's Financial Health," *Electrical World* 185 (March 15, 1976): 64.
76. As an exception, increased fuel costs sometimes could be passed along to customers as they were incurred.
77. Douglas D. Anderson, *Regulatory Politics and Electric Utilities: A Case Study in Political Economy* (Boston: Auburn House, 1981), pp. 70–1.
78. The problem of regulatory lag and other financial woes are discussed by utility officials and their critics in U.S. Congress, Senate, Committee on Interior and Insular Affairs, *Financial Problems of the Electric Utilities*, 93rd Congress, 2nd Session, Serial No. 93–50, August 7 and 8, 1974.
79. The grow-and-build strategy was definitely the paradigm at the TVA throughout most of the 1970s. Chapter 12 will describe the origins of the TVA conservation effort in 1978. Interview with James L. Williams, TVA director of Purchasing, January 19, 1984. As will be noted later, some utilities, such as New York City's Consolidated Edison Company, formally began promoting conservation rather than consumption in 1971.
80. Promotional tactics came under scrutiny by the Federal Power Commission and the House of Representatives in the late 1960s as utilities continued to offer incentives for contractors to install all-electric homes and offices. See Federal Power Commission, *Promotional Practices of Public Utilities: A Survey of Recent Actions by State Regulatory Commissions* (A Report to the Subcommittee on Regulatory Agencies, Select Committee on Small Business, U.S. House of Representatives) (Washington, DC: Federal Power Commission, 1970); "Dingell Unit Raps Utilities with Little Lady's Purse," *Electrical World* 170 (July 22, 1968): 17–18; "Dingell Committee Recommends States Regulate Incentives," *Electrical World* 170 (December 16, 1968): 24; and "Dingell Group Issues Report on Utilities' Promotion Practices," *Electrical World* 170 (November 11, 1968): 9. In 1970, the Virginia regulatory authorities prohibited the use of promotional allowances. "Virginia SCC tells Utilities to Stop Promotions," *Electrical World* 173 (May 11, 1970): 64. But at least one utility company in the state objected to the move. "Appalachian Power Fights Move to Kill Promotion Allowance," *Electrical World* 173 (March 30, 1970): 38. Promotional rate structures were disallowed in 1978 with passage of the Public Utility Regulatory Policies Act.
81. "Rate Structure May be the Next Target," *Electrical World* 177 (June 15, 1972): 7.
82. Interview with E. Floyd Thomas, former TVA manager of Power Operations, January 19, 1984. Data from Tennessee Valley Authority, *TVA 1982 Power Program Summary*, Vol. 1 (Knoxville, TN: Tennessee Valley Authority, 1983), p. 70. Electricity sales to all customers increased only 3% that year. *Ibid*.
83. This rate corresponds to a 2.8% annually compounded increase. Data from Edison Electric Institute, *Pocketbook*, p. 17.
84. Office of Technology Assessment, *New Electric Power Technologies: Problems and Prospects for the 1990s*, OTA-E-246 (Washington, DC: U.S. Congress, OTA, 1985), p. 45.

85. Interview with Burl B. Hulsey, Jr., vice chairman of the board, Texas Utilities, April 25, 1985.

86. This view was expressed by managers of the Madison Gas and Electric Company, August 21, 1984. These same managers were among the first to use the new concepts for developing novel rate structures in the 1970s, partly as a result of active "encouragement" by the Wisconsin regulatory authorities. The Energy Policy Staff of the President's Office of Science and Technology, directed by S. David Freeman, also noted in August 1970 that "the use of electricity is relatively inelastic with respect to price in the short run. . . ." Energy Policy Staff, Office of Science and Technology, *Electric Power and the Environment* (Washington, DC: Office of Science and Technology, 1970), p. 3; included in U.S. Congress, *Powerplant Siting*, p. 898. The Office of Technology Assessment report, *New Electric Power Technologies*, (on page 7) also notes how utilities "seriously underestimated the price elasticity of electricity demand" in the 1970s.

87. George C. Lodge and Irvin C. Bupp, "Report on the Conference About Electric Power for New England," Harvard Business School, January 31–February 2, 1975, p. 4. Draft report. Similar views about the inelasticity of electrical use came from an interview with Herman Koenig, former chairman of Michigan State University's Electrical Engineering Department, December 14, 1983.

88. Lodge and Bupp, "Report," p. 5.

89. Interview with Hugh Gene Hally, principal engineer, San Francisco Power Division, Bechtel Corporation, July 9, 1982.

90. Interview with S. David Freeman, director, TVA, January 17, 1984.

91. Interview with Robert Beckwith, System Planning manager, Commonwealth Edison Company, August 20, 1984.

92. Notes taken by George Wise, a General Electric Company historian, at the Utility Research Directors' Conference, March 29, 1976.

93. "Bagge Tries Confrontation, Chicago-style," *Electrical World* 173 (April 27, 1970): 7.

94. Carl E. Bagge, "The Electric Power Industry in Crisis and Transition: The Need for a National Energy-Environment Policy," *Proceedings of the American Power Conference* 32 (1970): 3.

95. *Ibid.*, p. 7.

96. Bagge reminisced about the 1970 confrontation in Carl E. Bagge, "The Snail Darter, Snapdragon, Saccharin Syndrome – More Power to Whom?" *Proceedings of the American Power Conference* 39 (1977): 10.

97. " 'Growth' Need Not be a Bad Word," *Electrical World* 173 (April 6, 1970): 9; and " 'Sheer Nonsense' Aymond Declares to Suggestion of a Power Crisis," *Edison Electric Institute Bulletin* 38 (1970): 7.

98. Inflation was viewed as the primary culprit in Carol J. Loomis, "For the Utilities, It's a Fight for Survival," *Fortune* 91 (March 1975): 189.

Chapter 10

1. "We're Already on the Way, Mr. Swidler," *Electrical World* 159 (June 17, 1963): 79.

2. *Ibid.*

3. Owen Ely, "FPC Chairman Swidler Favors Heavy R&D Program," *Public*

Utilities Fortnightly 72 (July 4, 1963): 44; and "What Others Think," *Public Utilities Fortnightly* 72 (July 4, 1963): 55.

4. Ely, "FPC Chairman," p. 46.

5. Swidler reiterated similar views when he compared R&D efforts in the electric utility industry with R&D performed by the telecommunications industry. "Testimony of Joseph C. Swidler," in *United States v. AT&T*, 1974, CA No. 74–1698, p. 9.

6. "Confident Mood, Fuss on Research Marks EEI Meeting," *Electrical World* 159 (June 17, 1963): 80–1.

7. Federal Power Commission, "Swidler Announces Formation of the (to be designated in the future) Ad Hoc Committee on Research and Development in the Electric Power Industry," Release No. 12838, July 30, 1963. Also cited in "As Electric Companies See It," *Electrical World* 161 (March 23, 1964): 130.

8. *Ibid.*, pp. 130–1.

9. *Ibid.*, p. 131.

10. "The Origin and Formation of the Electric Power Research Institute," a typed chronology of the history of the EPRI, found in the files of the Edison Electric Institute library, Washington, DC, no date. The name change was apparently announced in a letter of April 14, 1965 from the EEI to chief executive officers of EEI member companies.

11. Letter from Edwin Vennard, EEI Managing Director, to Chief Executives of EEI Member Companies, March 18, 1965, pp. 1–5 (located at EEI Library, Washington, DC); and EEI, *Report of the Ad Hoc Committee on Research and Development in the Electric Power Industry* (New York: Edison Electric Institute, March 15, 1965), p. 8. Also see "Utilities Team Up for Joint R&D Effort," *Electrical World* 163 (March 22, 1965): 87–8.

12. U.S. Congress, Senate, Committee on Commerce, *Electric Power Reliability*, Hearings, 90th Congress, 1st session, Serial No. 90–30, Part 1: August 22, 1967; Part 2: December 20 and 21, 1967; Part 3: April 26 and 29, 1968.

13. Federal Power Commission, *Prevention of Power Failures: An Analysis and Recommendations Pertaining to the Northeast Failure and the Reliability of U.S. Power Systems* (Washington, DC: U.S. Government Printing Office, 1967).

14. Several of these editorials are reprinted in U.S. Congress, "Electric Power Reliability," pp. 89–123.

15. "Swidler Again Prods Industry to Boost Research Efforts," *Electrical World* 173 (March 30, 1970): 19

16. Council on Economic Priorities, " *The Price of Power: Electric Utilities and the Environment* (New York: Council on Economic Priorities, 1972), pp. 25 and 31; included in U.S. Congress, Senate, Committee on Commerce, *Energy Research and Development*, Hearings, 92nd Congress, 1st session, Serial No. 92–62, March 15 and 16, 1972, pp. 259 and 265. Also see Federal Power Commission, *The 1970 National Power Survey*, Part 1 (Washington, DC: U.S. Government Printing Office, 1971), p. I-21-1. Put more bluntly by a trenchant utility critic, Democratic Senator Lee Metcalf of Montana, the annual R&D expenditures of several companies "would not pay the salary of a mediocre engineer for one summer." "FPC Rule Seeks to Encourage Utility R&D," *Electrical World* 173 (February 9, 1970): 7. Even Metcalf's charge could be viewed as an overstatement since several companies listed no R&D expenditures during 1969. U.S. Congress, *Energy Research and Development*, p. 149. Nevertheless, Metcalf reserved most of his invective for investor-owned utilities that continuously reduced their costs of providing electricity without reducing their rates to residential consumers by a proportional amount. See Lee Metcalf and Vic Reinemer, *Overcharge* (New York: David McKay, 1967). Some of Metcalf's criticisms can be read in "Metcalf Scores Utilities at Consumer Assembly '67," *Electrical World* 168 (November 13, 1967): 77–8.

17. Charles Erwin "Engine" Wilson served as president of General Motors from 1941 to 1953. Charles Edward "Electric" Wilson was president of General Electric from 1940 to 1942 and from 1944 to 1950. To avoid confusion between the two corporate presidents, the men took the nicknames "Engine" and "Electric." "Engine Charlie" died in 1961; "Electric Charlie" died in 1972. The origin of the nicknames is given in "Electric" Charlie's obituary in *The New York Times*, January 4, 1972, also found in *The New York Times Biographical Edition* 3 (January 1972): 240–1.

18. Letter from Philip Sporn, retired president, director, American Electric Power Company, to Senator John O. Pastore, chairman, Joint Committee on Atomic Energy, dated December 28, 1967, submitting his analysis entitled "Nuclear Power Economics – Analysis and Comments – 1967," contained in U.S. Congress, Joint Committee, *Nuclear Power Economics – 1962 Through 1967: Report of Joint Committee on Atomic Energy*, 90th Congress, 2nd Session (Washington, DC: U.S. Government Printing Office, 1968), pp. 16–17.

19. Carl E. Bagge, "The Electric Power Industry in Crisis and Transition: The Need for a National Energy-Environment Policy," *Proceedings of the American Power Conference* 32 (1970): 7.

20. "Powerplant Siting Act of 1971 – Amendment No. 364," *Congressional Record – Senate* (August 3, 1971): 28967–75. The amendment is also published in U.S. Congress, Senate, *Powerplant Siting*, 92nd Congress, 2nd Session (Washington, DC: U.S. Government Printing Office, 1972), pp. 29–38.

21. Text of Amendment to Senate Bill S.1684, "Title IV – Federal Power Research and Development Board Established," in U.S. Congress, *Energy Research and Development*, pp. 2–11.

22. *Ibid*., pp. 3–4 and 46.

23. Statement of Shearon Harris, chairman, Edison Electric Institute, in Senate Hearings, in U.S. Congress, *Energy Research and Development*, p. 78.

24. Statement of E. L. Kanouse, chairman, Electric Research Council, in U.S. Congress, *Energy Research and Development*, p. 77.

25. Statement of Shearon Harris, in U.S. Congress, *Energy Research and Development*, p. 91.

26. Press Release from the Electric Research Council, June 26, 1972. See also "The Electric Power Research Institute is Off and Running," *Public Utilities Fortnightly* 91 (May 10, 1973): 39–41. The article consists of a question and answer discussion with the new EPRI president Chauncey Starr.

27. Testimony by John P. Holdren, a Lawrence Livermore Laboratory physicist, in U.S. Congress, *Energy Research and Development*, pp. 155–6.

28. Statement of S. David Freeman, in U.S. Congress, *Energy Research and Development*, p. 45; and Electric Research Council, *Electric Utilities Industry Research and Development Goals Through the Year 2000: Report of the R&D Goals Task Force of the Electric Research Council* (New York: Electric Research Council, 1971), pp. 2–4.

29. Statement of Charles Luce, in U.S. Congress, *Energy Research and Development*, p. 99. Apparently, Luce had a change of heart on the matter of a federally funded and directed R&D center. In July 1970, he wrote a letter to Lee A.

DuBridge, science advisor to President Nixon, suggesting creation of a trust fund derived from an excise tax on electricity users. The fund, which would initially amount to $200 to $250 million per year, would be administered by a "national director" who received advice from the "utility industry, manufacturers, environmentalists, consumers, etc." He noted that while the Edison Electric Institute had been doing a "credible job" in raising funds for R&D from voluntary contributions of utilities, "the inevitable reluctance of companies to contribute to programs not supported by all segments of the industry . . . are severe handicaps to EEI's success in raising enough money." The letter of July 24, 1970 was ordered to be published along with other supporting evidence by Senator Magnuson and the Senate Committee on Commerce in the *Congressional Record – Senate* (August 3, 1971), p. 28969.

30. U.S. Congress, Senate, Committee on Commerce, *Energy Research and Development II*, 93rd Congress, 1st session, Serial No. 93–22, March 1, 1973, pp. 51–3.

31. Nilo Lindgren, "The First Five Years: Chauncey Starr and the Building of EPRI," *EPRI Journal* 3 (January/February 1978): 4–16; and Chauncey Starr, "The Electric Power Research Institute," *Science* 219 (1983): 1190–4.

32. Letter from Milton F. Searl, EPRI, to author, February 27, 1987.

33. See Electric Research Council, *Electric Utilities Research and Development Goals*, p. 5, for a criticism of how the utility industry's reliance on manufacturers for R&D would not be satisfactory in the future.

34. Press release from the Electric Research Council, June 26, 1972.

35. For details on Congressional action concerning energy legislation in 1973, see "Energy and Environment," *Congressional Quarterly Almanac* 29 (1973): 589–706.

36. Chauncey Starr, "Testimony on S.1283," in U.S. Congress, Senate, Committee on Interior and Insular Affairs, *Energy Research and Development Policy Act,* Hearings, 93rd Congress, 1st session, Serial No. 93–13, June 21 and 22, and July 11 and 12, 1973, pp. 132–3

37. *Ibid.*, p. 134.

38. *Ibid.*, pp. 136–7

39. "Statement of Joseph C. Swidler," in U.S. Congress, *Energy Research and Development Policy Act*, p. 91.

40. To be sure, the utility managers were not alone in their belief in technological progress. Even critics of the industry's behavior such as consumer advocate Ralph Nader reaffirmed his belief in the technological fix by arguing that innovation can be assured given the proper incentives and financial support. Testifying to Congress in 1973 about the industry's poor research efforts, he pointed out that: "We now have moved into a period of technological history, where we can almost . . . schedule innovation, we can almost program invention to solve the many technological deficiencies and by-products that now are so seriously affecting the nation." Testimony of Ralph Nader in U.S. Congress, *Research and Development II*, p. 27.

41. The utility industry's resistance to criticism can be explained from an organizational perspective as well. In general, people in organizations (bureaucracies, companies, industries, etc.) resent outsiders' interference in their affairs, believing that only they have the experience and expertise to provide useful information. This attitude makes it extremely difficult for "interlopers" to receive fair and evenhanded treatment of suggestions. Even if this were possible, it is nearly impossible to convince insiders to make radical changes. As one form of resistance, organizations find ways of discrediting critics and ensuring against fair evaluation of criticism not originating in the organization's power center. This certainly appears to have been the case in the utility industry during the 1960s and 1970s.

Chapter 11

1. Arthur M. Louis, "Southern California Edison Struggles to Bring Power to the People," *Fortune* 87 (May 1973): 214.
2. Among the many references concerning the power crisis that preceded the "energy crisis" of 1973 are: "Peak-shaving Aids Systems in PJM Area," *Electrical World* 167 (June 26, 1967): 45; "TVA Says Coal Supply May Run Out, Causing a Shortage of Power," *Wall Street Journal* (August 20, 1970), p. 28; "American Electric Says Its Supplies of Coal are 'Dangerously Low,' " *Wall Street Journal* (September 4, 1970), p. 8; "Gas Shortage Threatening to Worsen in Midwest and Northeast this Winter," *Wall Street Journal* (August 20, 1970), p. 28; "Coal Strap Tightens More in Southeast," *Electrical World* 173 (April 27, 1970): 22; "Panelist Says Power Rationing Already Here," *Electrical World* 173 (February 23, 1970): 21–2; "New York PSC chairman Joseph Swidler Predicts Brownouts This Summer," *Electrical World* 173 (March 9, 1970): 17; "White House Acts to Avert Summer Power Brownouts," *Electrical World* 173 (May 11, 1970): 33–4; " 'Power Crisis' to Spur Policy Change," *Electrical World* 173 (May 11, 1970): 38–41; and "New York Faces Brownouts; Con Edison in Worst Shape," *Electrical World* 173 (June 15, 1970): 18. Contemporary reviews of problems include "Why Utilities Can't Meet Demand," *Business Week*, No. 2100 (November 29, 1969): 48; Jeremy Main, "A Peak Load of Trouble for the Utilities," *Fortune* 80 (November 1969): 116–19, 194, 196, 200, and 205; and Lawrence A. Mayer, "Why the United States is in an Energy Crisis," *Fortune* 82 (November 1970): 75–7, 159–60, 162, and 164.
3. None of the bills passed Congress in 1970. See a review of the various hearings in "Electric Power Demands Clash with Environment," *Congressional Quarterly Almanac* 26 (1970): 523–5.
4. Comment by Shearon Harris, chairman of the Edison Electric Institute and chief executive of Carolina Power and Light, in "Energy Forum: 'Low-cost' Power Passe," *Electrical World* 176 (October 15, 1971): 25.
5. "Energy Conservation Hits Utilities in the Pocketbook," *Electrical World* 181 (March 15, 1974): 23–4.
6. Kenneth H. Bacon, "Rising Electricity Rates Prompt States, Cities to Consider Utility Take-Overs," *Wall Street Journal* (October 23, 1974), p. 38.
7. Sanford L. Jacobs, "Con Ed Omits Dividend for Second Quarter," *Wall Street Journal* (April 24, 1974), p. 3. Also see Alexander Lurkis, *The Power Brink: Con Edison – A Centennial of Electricity* (New York: Icare Press, 1982), pp. 121–4.
8. The company's stock traded for as much as 21.375 in early 1974. On April 22, it closed at 18; the next day at 12.25. On a day when total New York Stock Exchange volume amounted to just over 14 million shares, Con Ed stock was the most heavily traded, with 306,000 shares changing hands. Further selling of the stock brought the stock price down to 6.125 in July. Standard and Poor's Corporation, *Daily Stock Price Records, New York Stock Exchange*, Vols. 1 to 3 (New York: Standard and Poor's Corporation, 1974), pp. 91 and 92 in each volume. See also Victor J. Hillery, "Abreast of the Market," *Wall Street Journal* (April 24, 1974), p. 42, for a comment by an investment brokerage house vice president concerning the severity of the psychological impact of the Con Ed action.

9. Sanford L. Jacobs, "Con Ed Omits Dividend for Second Quarter," *Wall Street Journal* (April 24, 1974), p. 3.

10. Standard and Poor's Corporation, *Statistical Service Security Price Index Record*, 1982 Ed. (New York: Standard and Poor's Corporation, 1982), p. 96.

11. Standard and Poor's Corporation, *Trendline's Current Market Perspectives* 14 (January 14, 1975): 11.

12. U.S. Congress, Senate, Committee on Interior and Insular Affairs, *Financial Problems of the Electric Utilities*, 93rd Congress, 2nd session, Serial No. 93–50, August 7 and 8, 1974.

13. William G. Rosenberg, "Utilities Need Help – Now!" *Wall Street Journal* (January 8, 1975), p. 12. Mr. Rosenberg was chairman of the Michigan Public Service Commission and served as a member of the Project Independence Advisory Committee to the Federal Energy Agency. For another account of a utility's financial woes, see John Emshwiller, "How a Thriving Utility Became a Sagging One After its Giant Step," *Wall Street Journal* (October 8, 1974), p. 1. The article describes problems at Michigan's Consumers Power Company.

14. Swidler's criticism of the overconcentration of efforts on nuclear power began in the 1960s. See Owen Ely, "FPC chairman Swidler Favors Heavy R&D Program," *Public Utilities Fortnightly* 72 (July 4, 1963): 44; and interview with Joseph C. Swidler, former FPC chairman and New York State Public Service Commission chairman, March 15, 1983.

15. For almost two decades – beginning in the early 1950s – Sporn argued that utilities and manufacturers should not forego efforts to improve conventional fossil-fuel technology in order to pursue nuclear power. Standing alone (often without the support of natural allies such as the coal and gas industries), he frequently criticized the optimistic economic projects of the nuclear advocates and pointed out that fossil-fuel technology would still be needed into the 1970s and beyond. Research must be performed on metallurgy, he noted in 1957, so that steam temperatures could yield greater thermal efficiencies. Moreover, he advocated work on efforts to break the 40% thermal efficiency barrier with what became known as "combined-cycle" systems. Taking advantage of heat exhausted from a gas turbine generator, steam would be produced in a conventional power system, producing more electricity from each Btu of heat energy. Though the system offered excellent prospects for achieving its goal, "it has not received the attention it deserves." Philip Sporn, "Role of Research in Next Two Decades," paper given at annual conference for engineers, Ohio State University, May 3, 1957, in Philip Sporn, *Vistas in Electric Power*, Vol. 1 (Oxford: Pergamon Press, 1968), p. 234. In general, Sporn contended that conventional technology still held great promise – especially in regions of the country that had large deposits of fossil fuel – and that the government simply had been too inattentive to it by overemphasizing nuclear power. Irvin C. Bupp and Jean-Claude Derian, *The Failed Promise of Nuclear Power: The Story of Light Water* (New York: Basic Books, 1981), pp. 45–6.

16. Carl E. Bagge, "The Electric Power Industry in Crisis and Transition: The Need for a National Energy-Environment Policy," *Proceedings of the American Power Conference* 32 (1970): 6.

17. For a similar view, see D. C. Burnham, "Shifting to the Electric Economy," *Nuclear Energy Digest* 4 (1973): 1–2, 22–25. Mr. Burnham was chairman and chief executive officer of Westinghouse, a major manufacturer of nuclear equipment. *Nuclear Energy Digest* is a publication of the Westinghouse Electric Corporation.

18. George C. Lodge and Irvin C. Bupp, "Report on the Conference About Electric Power for New England," Harvard Business School, January 31–February 2, 1975, p. 5.
19. *Ibid*.
20. Carol J. Loomis, "For the Utilities, It's a Fight for Survival," *Fortune* 91 (March 1975): 97.

Chapter 12

1. "Increase in Sales, Higher Rates Help Restore Industry's Financial Health," *Electrical World* 185 (March 15, 1976): 64.
2. "The Climate's Bright, but with Clouds – Like Acid Rain," *Electrical World* 199 (April 1985): 71. At year's end, 185 rate cases with increases of $4.1 billion were still pending.
3. At $14.82 per month for 500 kilowatt-hours of electricity, they paid 44% more than the national average.
4. Federal Power Commission, *Typical Electric Bills* (Washington, DC: Federal Power Commission, 1969, 1971, and 1977). As already noted briefly, Con Ed's problems were highly complicated, resulting from conflicts with environmentalists concerning new construction sites, a strike at General Electric that delayed delivery of some equipment, and failures of its "Big Allis" unit and the nuclear power plant at Indian Point.
5. Thomas O'Hanlon, "Con Edison: The Company You Love to Hate," *Fortune* 73 (March 1966): 123–6, 170, and 173.
6. Peter Millones, "Con Ed Seeks Rise of $117.5 Million in Electric Rates," *The New York Times* (August 22, 1969), p. 1.
7. The deluge of complaints exceeded the 7,079 letters received during the entire year of 1970. Grace Lichtenstein, "State Adds Eight to Check on Con Ed," *The New York Times* (December 28, 1971), p. 21; and Grace Lichtenstein, "Two State Agencies are Investigating Con Ed Billing Practices," *The New York Times* (August 31, 1971), p. 37.
8. Incident cited in Douglas Anderson, *Regulatory Politics and Electric Utilities* (Boston: Auburn House, 1981), p. 68.
9. For more on lifeline rates, see *ibid*., Chapter 5, "The Politics of Pricing: Lifeline Rates and Regulatory Lag in California," pp. 135–65.
10. For examples of such testimony relating to bills that would have revised rate structures, see U.S. Congress, Senate, Committee on Commerce, *Electric Utility Rate Reform and Regulatory Improvement*, 94th Congress, 2nd Session, April 27 and 28, 1976, Serial No. 94–77; and U.S. Congress, Senate, Committee on Energy and Natural Resources, *Public Utility Rate Proposals of President Carter's Energy Program (Part E of S.1469)*, 95th Congress, 1st Session, July 27 and 28, and September 7, 8, 9, and 20, 1977, Serial No. 95–120, Parts 1, 2, and 3.
11. For example, the organization published its book, *How to Challenge Your Local Electric Utility* (by Richard Morgan and Sandra Jerabek) in 1974 and another (also authored by Richard Morgan) in 1980 called *The Rate Watcher's Guide: How to Shape Up Your Utility's Rate Structure*.
12. U.S. Congress, Senate, Committee on Commerce, *High Cost of Electricity*, 94th Congress, 2nd Session, October 7 and 8, 1976, Serial No. 94–106, p. 4.
13. For example, Eugene Daniell, a member of the New Hampshire House of

Representatives, testified about a supposed criminal conspiracy existing between the Public Service Company and a subsidiary of the Conoco Company that supplied the utility with coal. *Ibid*. p. 23.

14. *Ibid*., p. 45

15. Willliam C. Tallman, the utility company's president, submitted a statement that was included in the published hearings report. *Ibid*., pp. 198–204. According to an editorial published in the *Portsmouth (NH) Herald* on October 11, 1976, Senator Durkin gave representatives of the utility and PUC only one-day notice before convening what the paper called a "three-ring energy circus." "We Regret Our Support," *ibid*., p. 209.

16. See the description of the conflict between pollution and progress in Martin V. Melosi, *Coping with Abundance: Energy and Environment in Industrial America* (New York: Knopf, 1985), pp. 31–3. Despite protests by some people, pollution continued because "smoke was a visible sign of economic prosperity and material progress." *Ibid*., p. 33.

17. For a summary of the birth of the modern environmental movement, see Samuel P. Hays, *Beauty, Health, and Permanence: Environmental Politics in the United States, 1955–1985* (Cambridge: Cambridge University Press, 1987); and Melosi, *Coping with Abundance*, pp. 296–8. Also see summaries of environmental legislation in Congressional Quarterly, "Environmental Programs," *Congress and the Nation*, Vol. 3 (1969–72) (Washington, DC: Congressional Quarterly, 1973), pp. 745–7; and in subsequent volumes of this publication.

18. For a discussion of the antiscience and technology feelings during the 1960s and 1970s, see Daniel J. Kevles, *The Physicists: The History of a Scientific Community in Modern America* (New York: Knopf, 1978), pp. 393–409.

19. John Noble Wilford, "The Price of Technology," *The New York Times* (July 31, 1970), p. 34.

20. U.S. Congress, Senate, Committee on Government Operations, *Intergovenmental Coordination of Power Development and Environmental Protection Act*, 91st Congress, 2nd Session, February 3 and 4, April 29, and March 3, 1970, Part 1, p. 3.

21. Senator Kennedy summarized his earlier legislative attempts and their rationale in 1970, when he testified at a hearing for another bill. See U.S. Congress, Senate, Committee on Government Operations, *Intergovenmental Coordination of Power Development and Environmental Protection Act*, 91st Congress, 2nd Session, April 15, June 15, August 3, and September 16, 1970, Part 2, p. 520.

22. Quoted in "Why Utilities Can't Meet Demand," *Business Week*, No. 2100 (November 29, 1969): 56.

23. FPC commissioner Carl E. Bagge reprimanded utility managers for using this term in his 1970 American Power Conference talk. See Carl E. Bagge, "The Electric Power Industry in Crisis and Transition: The Need for a National Energy-Environment Policy," *Proceedings of the American Power Conference* 32 (1970): 3. In 1977, he further discouraged the use of the term "beautility." Carl E. Bagge, "The Snail Darter, Snapdragon, Saccharin Syndrome – More Power to Whom?" *Proceedings of the American Power Conference* 39 (1977): 9.

24. AEP, *Annual Report 1971*, p. 21. Electricity's role in cleaning up the environment may have been important, but most critics focused not on its utilization for these purposes – only the polluting side effects of its generation. Council on Economic Priorities, *The Price of Power: Electric Utilities and the Environment* (New

York: Council on Economic Priorities, 1972), p. 34, included in U.S. Congress, Senate, Committee on Commerce, *Energy Research and Development*, Hearings, 92nd Congress, 1st session, Serial No. 92–62, March 15 and 16, 1972, p. 268.

25. For example, see the comments of William H. Rodgers, a University of Washington law professor, in U.S. Congress, Senate, Committee on Congress, *Powerplant Siting*, 92nd Congress, 2nd Session, April 28, May 15, and June 1, 1972, Serial No. 92–92, pp. 149–61. Rodgers also wrote a critical book about the industry before the embargo. William Rodgers, *Brown-Out: The Power Crisis in America* (New York: Stein and Day, 1972).

26. U.S. Congress, Joint Committee on Atomic Energy, *Nuclear Power and Related Problems – 1968–1970*, 92nd Congress, 1st Session (Washington, DC: U.S. Government Printing Office, 1971), p. 68. Another major point of debate concerned whether the price of electricity (and other products) included the cost imposed on society to deal with the health, safety, and aesthetic effects of pollution. People such as Lee DuBridge, director of President Nixon's Office of Science and Technology, felt that it did not, and therefore electricity's price did not reflect its total cost on society. U.S. Congress, *Intergovernmental Coordination of Power Development and Environmental Protection Act*, Part 1, p. 41.

27. The new environment in which regulators tried to perform their jobs – and the implications of working in that environment – are discussed in Charles F. Phillips, Jr., *The Regulation of Public Utilities: Theory and Practice* (Arlington, VA: Public Utilities Reports, 1984), pp. 14–29. Not everyone had confidence in the regulatory bodies. In 1969, *The New York Times* called the New York State Public Service Commission "eminently unsuited to the task" of evaluating a rate request by Consolidated Edison. An upcoming decision, noted the editorial writer, would serve as a test of the commission's efforts to convince people that it really worked for them. See "Con Ed Wants More," *The New York Times* (August 22, 1969), p. 34.

28. Utility companies often argued that by seeking lower rates for consumers in the short run, regulators shortchanged ratepayers in the longer term because of the utilities' inability or unwillingness to build new power plants. Because of inadequate "rate relief," some utilities have taken the path of least resistance with a capital minimization strategy – a strategy that critics of regulatory policy say offers current benefits while promising a bleak future of capacity shortages, brownouts, and blackouts. See Peter Navarro, "Our Stake in the Electric Utility's Dilemma," *Harvard Business Review* 60 (May/June 1982): 87–97; and Peter Navarro, *The Dimming of America: The Real Costs of Electric Utility Regulatory Failure* (Cambridge, MA: Ballinger, 1985). Much testimony concerning the trend toward capital minimization strategies and how they may lead to capacity shortages in the future was given during an conference sponsored by the Federal Energy Regulatory Commission in 1981. See U.S. Federal Energy Regulatory Commission, "Informal Public Conference on the Financial Condition of the Electric Power Industry," Docket No. EL80–7-000, March 6, 1981.

29. Some customers of the Long Island Lighting Company paid twelve times more for electricity during peak times (such as during summer days) than on winter nights. For more on the evolution of this price structure in New York, see the chapter on Alfred Kahn (Chapter 7) in Thomas McCraw, *Prophets of Regulation* (Cambridge, MA: Harvard University Press, 1984). The author obtained other information on the history of this structure from an interview with managers of Madison Gas and Electric Company, August 21, 1984.

30. For a summary of the PURPA legislation, see "Energy Bill: The End of an Odyssey," in Congressional Quarterly, *1978 Congressional Quarterly Almanac* (Washington, DC: Congressional Quarterly, 1979), pp. 639–41 and 644–5. For an interpretation of how PURPA worked during its first few years, see Alvin L. Alm and Kathryn L. Stein, "PUPRA – Purpose and Prospects," Chapter 12 in James Plummer, Terry Ferrar, and William Hughes, eds., *Electric Power Strategic Issues* (Arlington, VA: Public Utilities Reports, 1983), pp. 237–64. Also see "California Orders its Utilities to 'Unsell' Energy," *Business Week*, No. 2638 (May 26, 1980): 167–76.
31. For an overview of the effect of siting regulations on utilities, see A. H. Ringleb, "Environmental Regulation of Electric Utilities," in John C. Moorhouse, ed., *Electric Power: Deregulation and the Public Interest* (San Francisco: Pacific Research Institute for Public Policy, 1986), pp. 190–5.
32. John V. Winter and David A. Conner, *Power Plant Siting* (New York: Van Nostrand Reinhold, 1978), p. 123.
33. *Ibid.*, pp. 40–5.
34. For a study of siting nuclear plants, see Joel Yellin and Paul L. Joskow, "Siting Nuclear Power Plants," *Virginia Journal of Natural Resources Law* 1 (1980): 1–67.
35. A. Joseph Dowd, "Environmental Considerations in Energy Production," *Public Utilities Fortnightly* 106 (October 9, 1980): 89.
36. O'Hanlon, "Con Edison: The Company You Love to Hate," p. 170.
37. Peter Millones, "Con Edison Challenges Physicist's Testimony on Nuclear Plant," *The New York Times* (June 15, 1967), p. 59.
38. Glenn Zorpette, "The Shoreham Saga," *IEEE Spectrum* 24 (November 1987): 33.
39. An outline of the PSC case against the Long Island Lighting Company can be found in J. Ronald Fox, "The Shoreham Nuclear Power Plant (A) and (B)," Cases 0–385–249 and 0–385–251, *Harvard Business School Case Studies Series* (1985).
40. Bill Paul, "Lilco Signs Pact with New York Over Shoreham," *Wall Street Journal* (June 17, 1988), p. 30.
41. "To Win Prudence, Woo the 'Stakeholder,' " *Electrical World* 202 (June 1988): 29–30.
42. Theodore J. Nagel, "Operating a Major Electric Utility Today," *Science* 201 (September 15, 1978): 991.
43. Emphasis added. Gordon C. Hurlbert, "The Anger of Decent Men," *Public Utilities Fortnightly* 103 (March 15, 1979): 16–19.
44. "Anatomy of a 'Murder Trial,' " *Electrical World* 181 (March 1, 1974): 13. A "review" article that did not look favorably on altering rate structures is "Rate Design: The Coming Nightmare?" *Electrical World* 184 (December 15, 1975): 71–4.
45. The analyst was John Attalienti of the Argus Research Corporation, cited in John Emshwiller, "Most Utilities Support Conservation With Much Talk but Little Action," *Wall Street Journal* (September 2, 1980), p. 29.
46. A commentary on the nature of prudence reviews, written by a former New York State Public Service commissioner, Alfred E. Kahn, is "Who Should Pay for Power-Plant Duds?" *Wall Street Journal* (August 15, 1985), p. 26. Also see Frederick Rose, " 'Prudency Reviews' Are Changing the Way Utilities Set Rates," *Wall Street Journal* (October 2, 1986), pp. 1 and 22; and Bill Paul, "More States Are Tightening Rules for Electric Utilities," *Wall Street Journal* (November 13, 1986), p. 6.
47. This view was presented to the author by Floyd L. Culler, president, Electric

Power Research Institute, January 16, 1986, and by several utility managers. It is also reflected in Vincent Butler, "A Social Compact to Be Restored," *Public Utilities Fortnightly* 116 (December 26, 1985): 17–21; and Irwin M. Stelzer, "The Utilities of the 1990s," *Wall Street Journal* (January 7, 1987), p. 20.

48. Lynn White, Jr., "Technology Assessment from the Stance of a Medieval Historian," *American Historical Review* 79 (1974): 3.

49. Even in the 1980s, some utility managers believed they were still in the business of selling electricity – not conservation. See Robert V. Percival, "Conservation and Renewable Energy Sources as Supply Alternatives for New York's Electric Utilities," in Sidney Saltzman and Richard E. Schuler, eds., *The Future of Electrical Energy: A Regional Perspective of an Industry in Transition* (New York: Praeger, 1986), p. 130.

50. Taking issue with FPC commissioner Carl Bagge's criticism of the utility industry's pursuit of nuclear power, A. H. Aymond, president of the Edison Electric Institute, argued along these lines as early as 1970. At the American Power Conference, where Bagge criticized the industry so forcefully, Aymond declared, "[a]n uninformed and misinformed public – that is the problem. We have failed to get them to understand that our technology will continue to advance. Not only will there be better means of controlling contaminants, but also new and better methods of harnessing energy. It is unrealistic and short-sighted to multiply what we are doing now by some factor and assume that this is what our children and grandchildren are going to be confronted with." " 'Power Crisis' to Spur Policy Change," *Electrical World* 173 (May 11, 1970): 38. A similar view about the public's lack of understanding of nuclear power was represented to the author by several utility managers.

51. See President's Commission on the Accident at Three Mile Island, *Report of the President's Commission on The Accident at Three Mile Island* (Washington, DC: U.S. Government Printing Office, 1979).

52. The company's system output increased 6.5% from 1969 to 1970. Pacific Gas and Electric Company, *Annual Report 1970*.

53. For an overview of load management techniques, see Office of Technology Assessment, *New Electric Power Technologies: Problems and Prospects for the 1990s*, OTA-E-246 (Washington, DC: U.S. Congress, OTA, 1985), pp. 142–3.

54. As noted earlier, utilities must construct sufficient facilities to meet peak demand, even if it is experienced for only a few minutes each day. Obviously, the cost of such construction can be very high and it can create financial problems for the company's customers, regulators, and investors. Therefore, if demand can be shifted from peak times to off-peak "valleys," the company can avoid building new facilities and mitigate associated problems. One incentive to customers consists of discounted "off-peak" pricing – a technique used by telephone companies to encourage use of facilities after most businesses have closed for the day and when usage has diminished significantly.

55. This discussion has been taken from David Roe, *Dynamos and Virgins* (New York: Random House, 1983) and reviews of the book by Richard F. Hirsh in *Technology Review* 88 (April 1985): 18; and *The Philadelphia Inquirer* (February 10, 1985), Book Review section, p. 6. A summary of the EDF approach can be found in John R. Emshwiller, "Environmental Group, In Change of Strategy, Is Stressing Economics," *Wall Street Journal* (September 28, 1981), pp. 1 and 23.

56. Also see Frederic D. Krupp, "New Environmentalism Factors In Economic

Needs," *Wall Street Journal* (November 20, 1986), p. 34. Mr. Krupp is the current executive director of the EDF.
57. "NEESplan First Update," NEES publication, 1980; and interviews of Joan T. Bok, current chairman, and Guy W. Nichols, former chairman of New England Electric System, May 8, 1985.
58. Comments by Samuel Huntington, president of New England Electric System, at Annual Meeting of Shareholders, April 23, 1985. Also see James Cook, "Living on Tenterhooks," *Forbes* 135 (June 17, 1985): 86–7.
59. See Charles Stein, "Utilities Stop Building, Bet on Low Electric Demand," *Boston Globe* (July 2, 1985), pp. 33 and 38.
60. The TVA, unlike other utilities, is a government-created organization and therefore has no stockholders. However, it is a self-financing entity, depending on funds generated by users and bondholders to provide the capital. In this case, then, the customers (along with bondholders) are also the investors, meaning that the management must be especially responsive to shifting values of its constituents.
61. William U. Chandler, *The Myth of TVA: Conservation and Development in the Tennessee Valley, 1933–1983* (Cambridge, MA: Ballinger, 1984), p. 140.
62. *Ibid.*, p. 147.
63. *Ibid.*, pp. 147–8.
64. Tennessee Valley Authority, *TVA 1982 Power Program Summary*, Vol. 1 (Knoxville, TN: Tennessee Valley Authority, 1983), p. 3.
65. Interview with S. David Freeman, director, TVA, January 17, 1984.
66. Chairman White was quoted at the 1981 AEP annual stockholders meeting. "AEP Planning for Future," *Roanoke Times and World News* (April 23, 1981), p. B-3.
67. Ernst R. Habricht, Jr., "Competition Can Power Utilities into the Black," *Wall Street Journal* (October 14, 1985), p. 14.
68. Utilities, however, cannot make too much money without causing regulatory commissions to require rate decreases and a distribution of some of their good fortune back to electricity rate-payers. An example of a recently diversifying company is the Wisconsin Electric Power Company, now called the Wisconsin Energy Company.
69. From 1982 through 1987, average real prices of electricity have been stable or declining. See U.S. Department of Energy, *Energy Security: A Report to the President of the United States* (Washington, DC: U.S. Department of Energy, 1987), p. 133.

Chapter 13

1. From 1979 through 1987, no new nuclear plants have been ordered in the United States. Some observers do not expect another nuclear plant to be ordered in the United States until the 1990s. See Peter Stoler, *Decline and Fail: The Ailing Nuclear Power Industry* (New York: Dodd, Mead, 1985).
2. This work was performed on the Harrison station units of the Monongahela Power Company, an operating company of the Allegheny Power System. According to Westinghouse engineers, the 50-Btu/kWh reduction presents a $10 million savings (at a rate in 1986 of $200,000/Btu). And at $2,000/kW – a modest cost for new capital expenditures – the 3.7-MW power increase becomes a savings of $7.4 million. Personal correspondence to author from J. David Conrad, manager, Business

Development, Power Generation, Westinghouse Electric Corporation, July 1986 and phone call, October 4, 1988.

3. Data from Westinghouse Steam Turbine-Generator Division, 1986.

4. Richard F. Hirsh, "Westinghouse Electric Corporation – Steam Turbine Division (A) and (B)," Cases 0–687–036 and 0–687–037, *Harvard Business School Case Studies Series* (1986). Also see "Westinghouse Improves Profit Margins; Concern Continues to Face Longer-Term Issues," *Wall Street Journal* (January 31, 1986), p. 6.

5. Correspondence to author from R. L. Evans, GE manager, Market Development, Power Plant Prospects, November 28, 1984. Also see Stephen C. Jenkins and Anthony C. Romania, "Facilities Life Extension: Broadening the View of Power Plant Upgrades," *Electric Forum* 9, No. 1 (1983): reprint pages 1–4. *Electric Forum* is a GE publication.

6. William C. Hayes, "Forecasting Calls for a Seer, Not an Analyst," *Electrical World* 200 (September 1986): 7. The view is resonated by Peter Blair, director of an Office of Technology Assessment project on new electric power technologies, who noted that conservation efforts have created "the most profound level of demand uncertainty in the history of electricity." See "After Oil: What Next?" *Newsweek* 107 (June 30, 1986): 68.

7. See John C. Sawhill and Lester P. Silverman, "Transformed Utilities: More Power to You," *Harvard Business Review* 63 (July-August 1985): 88–96.

8. "Toward Simplicity in Nuclear Plant Design," *EPRI Journal* 11 (July/August 1986): 8.

9. Paul L. Joskow and Richard Schmalensee, *Markets for Power: An Analysis of Electric Utility Deregulation* (Cambridge, MA: The MIT Press, 1983), pp. 51–2. The same conclusion is presented in Verne W. Loose and Theresa Flaim, "Economies of Scale and Reliability: The Economics of Large Versus Small Generating Units," *Energy Systems and Policy* 4 (1980): 37–56; and Kenneth Jameson, "Economies of Scale in the Electric Power Industry," in Kenneth Sayre, ed., *Values in the Electric Power Industry* (Notre Dame, IN: University of Notre Dame Press, 1977): 116–48; see particularly p. 135.

10. An Edison Electric Institute study of nuclear plants placed in service between 1973 and 1979 revealed that per-kilowatt capital costs were "generally independent of reactor size." Another study of French reactors in the size range between 900 and 1,300 MW also showed no economy of scale. "Toward Simplicity in Nuclear Design," p. 5. These studies challenge "long-held beliefs of the importance of economy of scale for nuclear plants." John J. Taylor, "The Opportunity for Small Reactors," *EPRI Journal* 11 (July/August 1986): 2.

11. Evan Herbert, "How the Electric Utilities Manage Cooperative R&D," *Research Management* 28 (September/October 1985): 16–17.

12. Four Japanese firms working on one type of fuel cell (the phosphoric acid fuel cell) consist of Mitsubishi Electric, Hitachi, Ltd., Toshiba Electric, and Fuji Electric companies. A description of their work is described in a presentation paper by Koji Kishida, "Status and Interest in Japanese Industrial Development of Fuel Cells," Mitsubishi Electric Corporation, Central Research Laboratory, 1985.

13. See, for example, A. P. Fickett, "Fuel Cell Power Plants," *Scientific American* 239 (December 1978): 70–6; Eliot Marshall, "The Procrastinator's Power Source," *Science* 224 (April 20, 1984): 268–70; and Ernest Raia, "Fuel Cells Spark Utilities' Interest," *High Technology* 5 (December 1984): 52–6. Also see Richard F. Hirsh,

"Prospects for Multi-Megawatt Fuel Cell Power Plants," Case 9–684–066, *Harvard Business School Case Studies Series* (revised 1985).

14. T. McCarthy, "Energy in 1980 and its Strategic Implications," internal document from General Electric Corporate Research Laboratories, February 5, 1960, p. 4.

15. Taylor Moore, "Utility Turbopower for the 1990s," *EPRI Journal* 13 (April/May 1988): 4–13.

16. For excellent descriptions of the IGCC plant and the interests that are working on it, see *New Electric Power Technologies: Problems and Prospects for the 1990s*, OTA-E-246 (Washington, DC: U.S. Congress, Office of Technology Assessment, 1985), pp. 108–12 and 272–7. For a more technical and economic analysis, see Seymour B. Alpert and Michael J. Gluckman, "Coal Gasification Systems for Power Generation," *Annual Review of Energy* 11 (1986): 315–55.

17. Much of this information comes from Taylor Moore, "How Advanced Options Stack Up," *EPRI Journal* 12 (July/August 1987): 4–13; and "Coal Gasification: Ready for the 1990s," *Electrical World* 201 (April 1987): 72–4.

18. According to EPRI, a technology in the pilot stage constitutes a concept verified by a small test facility. A demonstration plant is a concept farther along than a pilot plant verified by an integrated larger-scale plant. See EPRI, *TAG – Technical Assessment Guide. Vol. 1: Electricity Supply – 1986*, Report P-4463 SR (Palo Alto, CA: Electric Power Research Institute, 1986), p. B-27.

19. The Virginia Power and Potomac Electric Power companies have already committed themselves to combustion turbine installations with the option of adding combined-cycle units and coal gasifiers. See "Utilities Turning to Combustion Turbines to Meet Future Demand," *Electrical World* 201 (September 1987): 40–1. The Potomac Electric Power Company plans for a phased construction of a pair of 375-MW units, each consisting of two 125-MW combustion turbines and one 125-MW combined-cycle steam turbine, totalling 750 MW in one plant. See a description in "Pepco Plans GCC Capacity," *EPRI Journal* 12 (September 1987): 38.

20. See Westinghouse Electric Corporation promotional brochures: "Combustion Turbine Cogeneration" (1983) and "Take a New Look at Westinghouse Turbines" (July 1987).

21. See the entire issue of *Industrial Power Systems* 27 (December 1984), which is devoted to GE's efforts at selling cogeneration technology. (*Industrial Power Systems* is a GE publication.) GE also published several technical reports on cogeneration, such as W. G. Read, R. H. McMahan, Jr., and D. G. Wallace, "Cogeneration Financial Incentives," Report GER-3456 (1984).

22. "A 110-Megawatt Cogeneration Plant to Supply Steam to Pfizer's Citric Acid Plant While Generating Electricity for the Carolina Power and Light Company," *PR Newswire*, August 21, 1986, p. 82. *PR Newswire* can be obtained through the Dow Jones News Retrieval computer service. This story had an accession number of 04348999. Other information about Cogentrix and the Southport plant came from Cogentrix's 1987 *Annual Report* and promotional literature provided by the company. Its headquarters is in Charlotte, NC. For other examples of cogeneration successes, see Bill Paul, "Cogeneration is Rapidly Coming of Age," *Wall Street Journal* (March 2, 1987), p. 6.

23. Still more exotic technologies, such as fusion reactors and magnetohydrodynamic generators, continue to be discussed and supported by government agencies. These "revolutionary" technologies may not be available in small sizes, but they would overcome stasis by producing electricity in ways that sidestep the limits of thermodynamic cycles. These and other alternative technologies are described in Joseph J. DiCerto, *The Electric Wishing Well: The Solution to the Energy Crisis* (New York: Macmillan, 1976). An analysis of the likelihood of use of these technologies can be found in OTA, *New Electric Power Technologies*. For a description of the use of new materials that can withstand high temperatures in turbine blades (and other components), see Richard S. Claassen and Louis A. Girifalco, "Materials for Energy Utilization," *Scientific American* 255 (October 1986): 102–17. An editorial in *Science* also explores the prospects of cogeneration and combustion turbine technologies. Philip H. Abelson, "New Technologies in the Generation of Electricity," *Science* 236 (April 27, 1987): 373.

24. "Power Play," *The Economist* 307 (May 28, 1988): 20.

25. Interview with Arnold P. Fickett, EPRI department director, Energy Utilization, January 17, 1986.

26. Richard W. Zeren, "Adapting Energy Technology to the Future," *EPRI Journal* 11 (November 1986): 21.

27. Definitions for the work were: near-term, commercially available results within 10 years; intermediate, 10–25 years; long-term, 25 years or more. Contract expenditures in 1987 amounted to $259 million out of the total $325 million budget. Intermediate-term work received 34% of the $259 million. Long-term work received 5% of the expenditures. Data from EPRI, "Current Information," March 1987, p. 1.

28. Chauncey Starr, "The Electric Power Research Institute," *Science* 219 (1983): 1192.

29. Interview with Joseph C. Swidler, former FPC chairman and New York State Public Service Commission chairman, March 15, 1983.

30. For example, EPRI sponsors work (begun even before the organization's creation) on prototype systems for suspending coal particles on a bed of air in the boiler, permitting higher-temperature combustion. This "fluidized-bed" combustion technique increases thermal efficiency and breaks down impurities in the coal before they are exhausted as pollutants. It has been tested by the Tennessee Valley Authority (among others) and appears to be a promising technology for the 1990s. See Tennessee Valley Authority, *1982 Power Program Summary*, Vol. 1 (Knoxville, TN: Tennessee Valley Authority, 1983), p. 16. Also see S. David Freeman, "Remarks Before the Associated Press Managing Editors," Louisville, KY, November 2, 1983, pp. 8–9; Arnold M. Manaker and Patricia B. West, "TVA Orders 160 MWe Demonstration AFBC Power Station," *Modern Power Systems* (December-January 1984): 59; "AFBC Will Soon Be an Option, says EPRI Manager," *Electric Light and Power* 64 (May 1986): 24; and "How Fluidized Bed Technology Works," *Electrical World* 199 (June 1985): 106–7. Meanwhile, though the EPRI has been able to create expert teams to evaluate problems (such as the accident at Three Mile Island in 1979) and produce a vast amount of technical literature, Swidler noted that the research did not always lead to the introduction of new useful hardware – just reports. Interview with Swidler, March 15, 1983.

31. Data from U.S. Department of Energy, *Congressional Budget Request: FY 1986* (Washington, DC: United States Government Printing Office, 1985); included in OTA, *New Electric Power Technologies,* p. 255.

32. Stephen Breyer, *Regulation and Its Reform* (Cambridge, MA: Harvard University Press, 1982), p. 15.

33. PURPA was one of five laws that made up President Carter's energy plan, outlined to the public and Congress in April 1977. The goals of PURPA are listed in Section 101 of the act, Public Law 95–617, signed into law on November 9, 1978. A text of the law, along with explanatory statements from the House-Senate Conference Committee, can be found in U.S. Congress, Senate, *Senate Reports*, 95th Congress, 2nd Session, Vols. 1–13 ("Miscellaneous Reports on Public Bills, XIII"), Report No. 95–1292 (Washington, DC: U.S. Government Printing Office, 1978), pp. 1–121. The other laws that made up the Carter energy policy, also approved on November 9, 1978, were the "Energy Tax Act of 1978" (PL 95–618), the "National Energy Conservation Policy Act" (PL 95–619), the "Powerplant and Industrial Fuel Use Act of 1978" (PL 95–620), and the "Natural Gas Policy Act of 1978" (PL 95–621). President Carter's remarks made at the time of the bills' signing can be found in *Weekly Compilation of Presidential Documents* 14 (1978): 1978–85.
34. The Federal Energy Regulatory Commission (FERC), which was empowered to develop rules for implementing PURPA, ordered that utilities pay qualified cogenerators and small power producers the "full avoided cost" for their electricity. However, the determination of the incremental cost of a utility's power was made by state regulatory commissions. (See FERC's rules in 18 Code of Federal Regulations, part 292.) In some states, the avoided cost was defined simply as the utility's marginal cost of producing a kilowatt-hour, using only the cost of energy as the determinant. In other states, the avoided cost included an allowance for the capital cost of a plant that the utility does not need to build because of the existence of the cogenerator.
35. See Section 210(3)(1) of PURPA and 18 Code of Federal Regulations 292.601–602, which exempt qualified producers from financial and rate regulations contained in the Public Utility Holding Company Act of 1935 and certain state laws.
36. See sources on PURPA in Chapter 12 and Richard L. Gordon, *Reforming the Regulation of Electric Utilities: Priorities for the 1980s* (Lexington, MA: Lexington Books, 1982), pp. 155–8.
37. Thomas G. Donlan, "Power Play: Utilities Turn Up the Heat on Congeneration Rivals," *Barrons* 67 (November 23, 1987): 11, 59–61; "Is Cogeneration in for Hard Times?" *Electrical World* 21 (May 1987): 18; and "IPPs Had 4% of '86 Capacity," *Electrical World* 202 (September 1988): 23. According to another source, independent power producers only manufacture about 2.5% of the nation's electricity. Peter Nulty, "Utilities Flirt with Adam Smith," *Fortune* 117 (June 6, 1988): 176.
38. North American Electric Reliability Council, *1988 Reliability Assessment: The Future of Bulk Electric System Reliability in North America, 1988–1997* (Princeton: North American Electric Reliability Council, 1988), p. 12.
39. For a brief summary of FERC's experiment, see Richard L. Gordon, "Perspectives on Reforming Electric Utility Regulation," in John C. Moorhouse, ed., *Electric Power: Deregulation and the Public Interest* (San Francisco: Pacific Research Institute for Public Policy, 1986), p. 457.
40. As is admitted by the authors of these proposals, many of the implications of these deregulation models are not clear. Nevertheless, the idea is still gaining new supporters, and the literature on deregulation is growing rapidly. Some books and articles on the subject include Paul L. Joskow and Richard Schmalensee, *Markets for Power: An Analysis of Electric Utility Deregulation* (Cambridge, MA: The MIT Press, 1983); James Plummer, Terry Ferrar, and William Hughes, eds., *Electric*

Power Strategic Issues (Arlington, VA: Public Utilities Reports, 1983); Roger E. Bohn, Bennet W. Golub, Richard D. Tabors, and Fred C. Schweppe, "Deregulating the Generation of Electricity Through the Creation of Spot Markets for Bulk Power," *Energy Journal* 5 (April 1984): 71–91; Roger E. Bohn, "How to Coordinate and Value Dispersed Generation, Storage and End Use Technologies in Complex Power Systems," working paper, HBS 84–09, Harvard University, Graduate School of Business, 1983; Tom Alexander, "The Surge to Deregulate Electricity," *Fortune* 104 (July 13, 1981): 98–105; James L. Plummer, "A Different Approach to Electricity Deregulation," *Public Utilities Fortnightly* (July 7, 1983): 16–20; Carol E. Curtis, "Power Play," *Forbes* 134 (September 10, 1984): 160; "Competition to Shape the Industry, Utility Execs Hear," *Electrical World* 201 (July 1987): 17–18; and Ronald H. Schmidt, "Deregulating Electric Utilities: Issues and Implications," *Economic Review* (September 1987): 13–26. *Economic Review* is a publication of the Federal Reserve Bank of Dallas.

41. See Colin Norman, "Renewable Power Sparks Financial Interest," *Science* 212 (June 26, 1981): 1479–81.

42. *Federal Energy Regulatory Commission v. Mississippi*, 102 S.Ct. 2126 (1982); and *American Paper Institute v. American Electric Power Service Corporation*, 103 S.Ct. 1921 (1983). The latter case firmly (and finally) established the legitimacy of PURPA and subsequent FERC rules mandating that full avoided costs were to be paid by utilities to qualified cogenerators and small power producers. It also overcame objections of utility companies concerning the requirement to make interconnections between the producers and utility transmission grids. The case can be found in *Supreme Court Reporter* (St. Paul, MN: West Publishing Company, 1986), pp. 1921–33. A summary and brief discussion of the the case's implications was published as "High Court Upholds Utility Rules of United States," *The New York Times* (May 17, 1983), p. D5. The continuing controversy between regulated utilities and the new "PURPA producers" is portrayed in U.S. Congress, Senate, Committee on Energy and Natural Resources, *Implementation of the Public Utility Regulatory Policies Act of 1978*, 99th Congress, 2nd Session, S. Hearing 99–820, June 2 and 5, 1986.

43. "Is Cogeneration in for Hard Times?" p. 18.

44. William C. Hayes, "What Level of Competition Can We Afford?" *Electrical World* 201 (August 1987): 7. For other criticisms of PURPA, see George Melloan, "Californians Will Pay Dearly for PURPA Power," *Wall Street Journal* (March 31, 1987), p. 37. In the late 1980s, the situation has arisen in several states in which independent companies want to produce more power than is needed for the foreseeable future. As a result, state regulatory bodies have been studying and implementing ways to avoid creation of excess capacity. In Virginia, for example, the state regulatory commission has investigated a competitive bidding scheme for unregulated companies that wish to sell power to regulated utilities. See Commonwealth of Virginia, State Corporation Commission, "Staff Report, Case No. PUE870080," December 4, 1987.

45. William C. Hayes, "You can use it, but . . . ," *Electrical World* 201 (September 1987): 7.

46. See Stuart Diamond, "Study Cites Shift in Power Plants: More Use of Alternate Energy Threatens Dominance of Utilities, Report Says," *The New York Times* (November 25, 1984), p. 37.

47. Data from Alvin L. Alm and Kathryn L. Stein, "PURPA – Purpose and

Prospects," Chapter 12 in James Plummer, Terry Ferrar, and William Hughes, *Electric Power Strategic Issues* (Arlington, VA: Public Utilities Reports, 1983), pp. 241 and 246.

48. Melloan, "Californians Will Pay Dearly for PURPA Power," p. 37. Another impressive "joiner" in the cogeneration movement is the Southern California Edison Company, which in 1987 had negotiated contracts for 4,000 MW of cogenerated power and was already receiving 800 MW. See "1987 to Spotlight Competition and Deregulation," *Electrical World* 201 (January 1987): 15.

49. See Berry's remarks before the Edison Electric Institute, Fall Financial Conference, Palm Beach, Florida, October 6, 1981, "Let's End the Monopoly," reproduced and distributed by VEPCo. Also see William W. Berry, "The Deregulated Electric Utility Industry," 1981, obtained from VEPCo; and William Berry, "The Deregulated Electric Utility Industry," in Plummer, *Electric Power Strategic Issues*, pp. 3–25.

50. James T. Evans, "Generating Competition," *Powerline* (Fall 1987): 5. *Powerline* is a publication of Virginia Power Company.

51. "Statement of William W. Berry, chairman of the board, Virginia Power, before the Federal Energy Regulatory Commission Concerning Implementation of the Public Utility Regulatory Policies Act," Docket Rm 87–12–000, Washington, DC, April 16, 1987, p. 5.

52. Janet Novack, "The Regulatory Sidestep," *Forbes* 141 (May 2, 1988): 77.

Chapter 14

1. Philip Sporn, "Growth and Development in the Electric Power Industry," *Electrical Engineering* 78 (May 1959): 555.

2. A strong case for using history in policy making can be found in Thomas P. Hughes, "Technological History and Technical Problems," Chapter 12 in Chauncey Starr and Philip C. Ritterbush, eds., *Science, Technology, and the Human Prospect* (New York: Pergamon Press, 1980), pp. 141–156.

3. Michael A. Saren and Douglas T. Brownlie, *A Review of Technology Forecasting Techniques and Their Application* (Bradford [Yorkshire]: MCB University Press, 1983), p. 14. See also Steven C. Wheelwright and Spyros Makridakis, *Forecasting Methods for Management*, 4th Ed., especially Chapter 13 on "Qualitative and Technological Approaches to Forecasting" (New York: Wiley, 1985).

4. Interview with Eugene J. Cattabiani, executive vice president, Power Generation, Westinghouse Electric Corporation, July 29, 1985.

5. "NEES: From Industry Rogue to Guiding Light," *The Energy Daily* (November 20, 1981), p. 3.

6. In 1967, Luce replaced Harland C. Forbes – an "aggressive engineer" who wanted to place a nuclear plant in the heart of New York City. Luce had a background in law, government, and public relations and had previously served as undersecretary of the U.S. Department of Interior and head of the Bonneville Power Administration. The appointment of Mr. Luce was a "move that stunned the industry" according to *Electrical World*. "Ex-BPA Chief will Head Con Edison," *Electrical World* 167 (April 3, 1967): 55. An interview of the new CEO can be found in "Can Luce Give Con Edison a New Image?" *Electrical World* 168 (October 23, 1967): 31–5. For the details of the attempt to build a nuclear plant in New York

City, see George T. Mazuzan, " 'Very Risky Business': A Power Reactor for New York City," *Technology and Culture* 27 (1986): 262–84.

7. "Con Edison Seeks 10% Load Cut Now, Maybe 10% More Later; Luce: 'It's Sanity,' " *Electrical World* 175 (June 1, 1971): 30.

8. Perhaps the strategy worked well only in the long run for Con Ed. In the short run, reduced electricity sales meant lower revenues, which only added to the company's problems. One reason for the company's omission of a dividend in 1974, after all, was because of what the company declared was a "severe cash shortage and a persistent decline in sales." The company's cash position worsened as fuel prices soared and as "customers, stung by bigger bills despite efforts to use less electricity, have paid bills more slowly, or withheld payment altogether." Sanford L. Jacobs, "Con Ed Omits Dividend for Second Quarter," *Wall Street Journal* (April 24, 1974), p. 3. Con Ed, of course, did not go bankrupt. Actually, the company emerged from its status as the "basket case among utilities" in the late 1970s to become financially viable again. Harold Seneker, "Con Edison: Riches to Rags and Back to Riches Again," *Forbes* 122 (July 10, 1978): 56–8.

9. Among the many publications that warn of capacity shortages in the 1990s are: Peter Navarro, *The Dimming of America* (Cambridge, MA: Ballinger Press, 1985); Bill Paul, "Utilities Say This Summer's Brownouts Will Escalate to Severe Shortages in 1990s," *Wall Street Journal* (June 17, 1985), p. 30; and Mark Crawford, "The Electricity Industry's Dilemma," *Science* 229 (July 19, 1985): 248–50. Also see North American Electric Reliability Council (NERC), *1986 Reliability Review: A Review of Bulk Power System Reliability in North America* (Princeton: North American Reliability Council, 1986), which seriously questions the ability of utilities to provide enough capacity for even a 2.2% annual growth rate in demand between 1986 and 1995. A similar view is argued by Jerry Geist, chairman of the Edison Electric Institute – a utility trade organization. See "The Final Stakes are High in the Capacity Crapshoot," *Electrical World* 201 (March 1987): 15–16.

10. For other specific pieces of advice based on the experience of the utility industry, see Richard F. Hirsh, "How Success Short-Circuits the Future," *Harvard Business Review* 64 (March-April 1986): 72–6. Unfortunately, even this article does not contain what a manager really would want – a "stasis meter" that would indicate whether an individual industry or company is approaching stasis. From a study of just one industry, it is unlikely that a reliable scale could be designed. Moreover, from a theoretical point of view, such a meter might not be possible because of the difficulty (impossibility?) of quantifying terms (such as technological innovation) and because of problems in assessing the potential impact of new technical developments.

Appendix A

1. Thomas S. Kuhn, *The Structure of Scientific Revolutions*, 2nd. Ed. (Chicago: University of Chicago Press, 1970).

2. Edward W. Constant II, *The Origins of the Turbojet Revolution* (Baltimore: Johns Hopkins University Press, 1980).

3. Arthur L. Robinson, "Problems with Ultraminiaturized Transistors," *Science* 208 (June 13, 1980): 1246–9.

4. Louis T. Wells, Jr., *Product Life Cycle and International Trade* (Boston: Harvard Graduate School of Business Administration, Division of Research, 1972).

5. Chauncey Starr and Richard Rudman, "Parameters of Technological Growth," *Science* 182 (October 26, 1973): 358–64.
6. William J. Abernathy, Kim B. Clark, and Alan M. Kantrow, *Industrial Renaissance: Producing a Competitive Future for America* (New York: Basic Books, 1983).
7. *Ibid.*, pp. 17–18.
8. *Ibid.*, p. 18.
9. See also William J. Abernathy and Kenneth Wayne, "Limits of the Learning Curve," *Harvard Business Review* article republished in Alan M. Kantrow and Richard S. Rosenbloom, *The Management of Technological Innovation* (Boston: Harvard Business Review, 1982), pp. 55–65.
10. Abernathy, *Industrial Renaissance*, p. 15.
11. See also William J. Abernathy and Phillip L. Townsend, "Technology, Productivity and Process Change," *Technological Forecasting and Social Change* 7 (1975): 379–96.
12. Abernathy, *Industrial Renaissance*, p. 22. This is somewhat similar to Constant's notion of coevolution in design.
13. *Ibid.*, p. 24.
14. *Ibid.*, p. 27.
15. According to John Staudenmaier, the people working on maintaining and extending the extant technology are the "maintenance constituency." These are people (or groups and institutions) who have come to depend on the design and who profit from its maintenance. See John M. Staudenmaier, *Technology's Storytellers: Reweaving the Human Fabric* (Cambridge, MA: The MIT Press, 1985), pp. 195–6. As has already been discussed, these people have developed a value system around the technological system and are therefore likely to continue to push its development further in subsequent years, despite evidence that might suggest its lack of vitality.
16. Abernathy, *Industrial Renaissance*, p. 20.
17. For other studies on models of technological innovation, see William J. Abernathy and Kim B. Clark, "Innovation: Mapping the Winds of Creative Destruction," *Research Policy* 14 (1985): 3–22. This paper offers a discussion of the "transilience matrix" for understanding innovative events and their impacts in industries. Also see Chris DeBresson and Joseph Lampel, "Beyond the Life Cycle: Organizational and Technological Design. I. An Alternative Perspective," *Journal of Productivity and Innovation Management* 3 (1985): 170–87 and the following article by the same authors, "Beyond the Life Cycle. II. An Illustration," *ibid.*, pp. 188–95. For discussions on the use and misuse of standard technology life-cycle models, see Michael A. Saren and Douglas T. Brownlie, *A Review of Technology Forecasting Techniques and Their Application* (Bradford [Yorkshire]: MCB University Press, 1983); and Steven C. Wheelwright and Spyros Makridakis, *Forecasting Methods for Management*, 4th Ed. (New York: Wiley, 1985).
18. Staudenmaier, *Technology's Storytellers*, p. 197.
19. The discipline of history of technology is one of the few fields that has paid special attention to the importance of national "styles" in technological development. Such an emphasis is one of the virtues of Thomas P. Hughes, *Networks of Power: Electrification in Western Society, 1880–1930* (Baltimore: Johns Hopkins University Press, 1983).
20. Comments made by Chauncey Starr, former president of the Electric Power Research Institute and currently its vice chairman, January 16, 1986.

21. Arthur A. Bright, *The Electric Lamp Industry* (New York: Macmillan, 1949), pp. 126–7, cited in Leonard S. Reich, *The Making of American Industrial Research: Science and Business at GE and Bell, 1876–1926* (Cambridge: Cambridge University Press, 1985), pp. 62–3.
22. They already are, but because of past technical problems and popular distaste for the current form of light-water and boiling-water reactors, work has begun on "inherently safe" nuclear reactors. See Russ Manning, "PIUS Holds Promise for Nuclear's Future," *Electrical World* 198 (November 1984): 73–5. PIUS is an acronym for Process Inherent Ultimate Safety. Also see Richard K. Lester, "Rethinking Nuclear Power," *Scientific American* 254 (March 1986): 31–9.

Appendix B

1. Slightly modified from drawing in W. D. Marsh, *Economics of Electric Utility Power Generation* (Oxford: Clarendon Press, 1980), p. 101.
2. R. C. Spencer, "Evolution in the Design of Large Steam Turbine-Generators," General Electric Large Steam Turbine Seminar, Paper 83T1, 1983, p. 4.
3. Interview with Eugene J. Cattabiani, executive vice president, Power Generation Division, Westinghouse Electric Corporation, July 29, 1985.
4. G. O. Wessenauer, W. E. Dean, Jr., and J. E. Gilleland, "Some Problems on a Large Power System with Large Generating Units." Paper presented at the Sixth World Power Conference, Melbourne, Australia, October 20–27, 1962, p. 5.
5. See John H. DeYoung, Jr., and John E. Tilton, *Public Policy and the Diffusion of Technology: An International Comparison of Large Fossil-Fueled Generating Units* (University Park, PA: Pennsylvania State University Press, 1978), pp. 8–11.
6. More precisely, the freezing point of water is 491.69 °R. So 0 °F equals about 460 °R.
7. Much of this discussion comes from an excellent chapter on the physics of thermal efficiency contained in Paul H. Coortner and George O. G. Lof, *Water Demand for Steam Electric Generation: An Economic Projection Model* (Washington, DC: Resources for the Future, 1965 [distributed by Johns Hopkins Press]), pp. 87–120.
8. "Conventional Steam-Electric Generating Stations," Advisory Committee Report No. 7, April 1963, Part 2, p. 52 of U.S. Federal Power Commission, *National Power Survey* (Washington, DC: U.S. Government Printing Office, 1964).

Index

A

B

C

D

E

F

R

S

T

U

V

W

Z

Printed in the United States
By Bookmasters